BEN GOLDACRE is a writer, broadcaster and doctor best known for the 'Bad Science' column in the *Guardian*. Trained in Oxford and London, with brief forays into academia, he is thirty-four and works full-time for the NHS.

BEN GOLDACRE

Bad Science

FOURTH ESTATE · *London*

First published in Great Britain in 2008 by
Fourth Estate
An imprint of HarperCollins*Publishers*
77–85 Fulham Palace Road,
Hammersmith, London W6 8JB
www.4thestate.co.uk

Visit our authors' blog: www.fifthestate.co.uk

Diagrams © HarperCollins*Publishers*, designed by
HL Studios, Oxfordshire

2

A catalogue record for this book is available from the British Library

ISBN 978-0-00-724019-7

Typeset in Minion with Helvetica display by
G&M Designs Limited, Raunds, Northamptonshire

Printed and bound in Great Britain by Clays Ltd, St Ives plc

Mixed Sources
Product group from well-managed
forests and other controlled sources
www.fsc.org Cert no. SW-COC-1806
© 1996 Forest Stewardship Council

FSC

FSC is a non-profit international organisation established to promote the
responsible management of the world's forests. Products carrying the FSC
label are independently certified to assure customers that they come
from forests that are managed to meet the social, economic and
ecological needs of present and future generations.

Find out more about HarperCollins and the environment at
www.harpercollins.co.uk/green

To whom it may concern

CONTENTS

INTRODUCTION

Let me tell you how bad things have become. Children are routinely being taught – by their own teachers, in thousands of British state schools – that if they wiggle their head up and down it will increase blood flow to the frontal lobes, thus improving concentration; that rubbing their fingers together in a special sciencey way will improve 'energy flow' through the body; that there is no water in processed food; and that holding water on their tongue will hydrate the brain directly through the roof of the mouth, all as part of a special exercise programme called 'Brain Gym'. We will devote some time to these beliefs and, more importantly, the buffoons in our education system who endorse them.

But this book is not a collection of trivial absurdities. It follows a natural crescendo, from the foolishness of quacks, via the credence they are given in the mainstream media, through the tricks of the £30 billion food supplements industry, the evils of the £300 billion pharmaceuticals industry, the tragedy of science reporting, and on to cases where people have wound up in prison, derided, or dead, simply through the poor understanding of statistics and evidence that pervades our society.

At the time of C.P. Snow's famous lecture on the 'Two Cultures' of science and the humanities half a century ago, arts graduates simply ignored us. Today, scientists and doctors find themselves outnumbered and outgunned by vast armies of individuals who feel entitled to pass judgement on matters of evidence – an admirable aspiration – without troubling themselves to obtain a basic understanding of the issues.

At school you were taught about chemicals in test tubes, equations to describe motion, and maybe something on photosynthesis – about which more later – but in all likelihood you were taught nothing about death, risk, statistics, and the science of what will kill or cure you. The hole in our culture is gaping: evidence-based medicine, the ultimate applied science, contains some of the cleverest ideas from the past two centuries, it has saved millions of lives, but there has never once been a single exhibit on the subject in London's Science Museum.

This is not for a lack of interest. We are obsessed with health – half of all science stories in the media are medical – and are repeatedly bombarded with sciencey-sounding claims and stories. But as you will see, we get our information from the very people who have repeatedly demonstrated themselves to be incapable of reading, interpreting and bearing reliable witness to the scientific evidence.

Before we get started, let me map out the territory.

Firstly, we will look at what it means to do an experiment, to see the results with your own eyes, and judge whether they fit with a given theory, or whether an alternative is more compelling. You may find these early steps childish and patronising – the examples are certainly refreshingly absurd – but they have all been promoted credulously and with great authority in the mainstream media. We will look at the attraction of sciencey-sounding stories about our bodies, and the confusion they can cause.

Then we will move on to homeopathy, not because it's important or dangerous – it's not – but because it is the perfect model for teaching evidence-based medicine: homeopathy pills are, after all, empty little sugar pills which seem to work, and so they embody everything you need to know about 'fair tests' of a treatment, and how we can be misled into thinking that any intervention is more effective than it really is. You will learn all there is to know about how to do a trial properly, and how to spot a bad one. Hiding in the background is the placebo effect, probably the most fascinating and misunderstood aspect of human healing, which goes far beyond a mere sugar pill: it is counterintuitive, it is strange, it is the true story of mind–body healing, and it is far more interesting than any made-up nonsense about therapeutic quantum energy patterns. We will review the evidence on its power, and you will draw your own conclusions.

Then we move on to the bigger fish. Nutritionists are alternative therapists, but have somehow managed to brand themselves as men and women of science. Their errors are much more interesting than those of the homeopaths, because they have a grain of real science to them, and that makes them not only more interesting, but also more dangerous, because the real threat from cranks is not that their customers might die – there is the odd case, although it seems crass to harp on about them – but that they systematically undermine the public's understanding of the very nature of evidence.

We will see the rhetorical sleights of hand and amateurish errors that have led to you being repeatedly misled about food and nutrition, and how this new industry acts as a distraction from the genuine lifestyle risk factors for ill health, as well as its more subtle but equally alarming impact on the way we see ourselves and our bodies, specifically in the widespread move to medicalise social and political problems, to conceive of them in a reductionist, biomedical framework, and peddle commodifiable solutions, particularly in the form of pills and faddish diets. I will show you evidence that a vanguard of startling wrongness is entering British universities, alongside genuine academic research into nutrition. This is also the section where you will find the nation's favourite doctor, Gillian McKeith, PhD. Then we apply these same tools to proper medicine, and see the tricks used by the pharmaceutical industry to pull the wool over the eyes of doctors and patients.

Next we will examine how the media promote the public misunderstanding of science, their single-minded passion for pointless non-stories, and their basic misunderstandings of statistics and evidence, which illustrate the very core of why we do science: to prevent ourselves from being misled by our own atomised experiences and prejudices. Finally, in the part of the book I find most worrying, we will see how people in positions of great power, who should know better, still commit basic errors, with grave consequences; and we will see how the media's cynical distortion of evidence in two specific health scares reached dangerous and frankly grotesque extremes. It's your job to notice, as we go, how incredibly prevalent this stuff is, but also to think what you might do about it.

You cannot reason people out of positions they didn't reason themselves into. But by the end of this book you'll have the tools to win – or at least understand – any argument you choose to initiate, whether it's on miracle cures, MMR, the evils of big pharma, the likelihood of a given vegetable preventing cancer, the dumbing down of science reporting, dubious health scares, the merits of anecdotal evidence, the relationship between body and mind, the science of irrationality, the medicalisation of everyday life, and more. You'll have seen the evidence behind some very popular deceptions, but along the way you'll also have picked up everything useful there is to know about research, levels of evidence, bias, statistics (relax), the history of science, anti-science movements and quackery, as well as falling over just some of the amazing stories that the natural sciences can tell us about the world along the way.

It won't be even slightly difficult, because this is the only science lesson where I can guarantee that the people making the stupid mistakes won't be you. And if, by the end, you reckon you might still disagree with me, then I offer you this: you'll still be wrong, but you'll be wrong with a lot more panache and flair than you could possibly manage right now.

Ben Goldacre
July 2008

1

Matter

I spend a lot of time talking to people who disagree with me – I would go so far as to say that it's my favourite leisure activity – and repeatedly I meet individuals who are eager to share their views on science despite the fact that they have *never done an experiment*. They have never tested an idea for themselves, using their own hands; or seen the results of that test, using their own eyes; and they have never thought carefully about what those results mean for the idea they are testing, using their own brain. To these people 'science' is a monolith, a mystery, and an authority, rather than a method.

Dismantling our early, more outrageous pseudoscientific claims is an excellent way to learn the basics of science, partly because science is largely about disproving theories, but also because the lack of scientific knowledge among miracle-cure artistes, marketers and journalists gives us some very simple ideas to test. Their knowledge of science is rudimentary, so as well as making basic errors of reasoning, they also rely on notions like magnetism, oxygen, water, 'energy' and toxins: ideas from GCSE-level science, and all very much within the realm of kitchen chemistry.

Detox and the theatre of goo

Since you'll want your first experiment to be authentically messy, we'll start with detox. Aqua Detox is a detox footbath, one of many similar products. It has been promoted uncritically in some very embarrassing articles in the *Telegraph*, the *Mirror*, the *Sunday Times*, *GQ* magazine and various TV shows. Here is a taster from the *Mirror*.

> We sent Alex for a new treatment called Aqua Detox which releases toxins before your eyes. Alex says: 'I place my feet in a bowl of water, while therapist Mirka pours salt drops in an ionising unit, which will adjust the bio-energetic field of the water and encourage my body to discharge toxins. The water changes colour as the toxins are released. After half an hour, the water's turned red … she gets our photographer Karen to give it a go. She gets a bowl of brown bubbles. Mirka diagnoses an overloaded liver and lymph – Karen needs to drink less alcohol and more water. Wow, I feel virtuous!'

The hypothesis from these companies is very clear: your body is full of 'toxins', whatever those may be; your feet are filled with special 'pores' (discovered by ancient Chinese scientists, no less); you put your feet in the bath, the toxins are extracted, and the water goes brown. Is the brown in the water because of the toxins? Or is that merely theatre?

One way to test this is to go along and have an Aqua Detox treatment yourself at a health spa, beauty salon, or any of the thousands of places they are available online, and take your feet out of the bath when the therapist leaves the room. If the water goes brown without your feet in it, then it wasn't your feet or your toxins that did it. That is a controlled experiment: everything is the same in both conditions, except for the presence or absence of your feet.

There are disadvantages with this experimental method (and there is an important lesson here, that we must often weigh up the benefits and practicalities of different forms of research, which will become important in later chapters). From a practical perspective, the 'feet out' experiment involves subterfuge, which may make you uncomfortable. But it is also expensive: one session of Aqua Detox will cost more than the components to build your own detox device, a perfect model of the real one.

You will need:
- One car battery charger
- Two large nails
- Kitchen salt
- Warm water
- One Barbie doll
- A full analytic laboratory (optional)

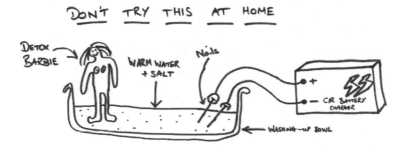

This experiment involves electricity and water. In a world of hurricane hunters and volcanologists, we must accept that everyone sets their own level of risk tolerance. You might well give yourself a nasty electric shock if you perform this experiment at home, and it could easily blow the wiring in your house. It is not safe, but it is in some sense relevant to your understanding of MMR, homeopathy, post-modernist critiques of science and the evils of big pharma. Do not build it.

When you switch your Barbie Detox machine on, you will see that the water goes brown, due to a very simple process called electrolysis: the iron electrodes rust, essentially, and the brown rust goes into the water. But there is something more happening in there, something you might half-remember from chemistry at school. There is salt in the water. The proper scientific term for household salt is 'sodium chloride': in solution, this means that there are chloride ions floating around, which have a negative charge (and sodium ions, which have a positive charge). The red connector on your car battery charger is a 'positive electrode', and here, negatively charged electrons are stolen away from the negatively charged chloride ions, resulting in the production of free chlorine gas.

So chlorine gas is given off by the Barbie Detox bath, and indeed by the Aqua Detox footbath; and the people who use this product have elegantly woven that distinctive chlorine aroma into their story: it's the chemicals, they explain; it's the chlorine coming out of your body, from all the plastic packaging on your food, and all those years bathing in chemical swimming pools. 'It has been interesting to see the colour of the water change and smell the chlorine leaving my body,' says one testimonial for the similar product Emerald Detox. At another sales site: 'The first time she tried the Q2 [Energy Spa], her business partner said his eyes were burning from all the chlorine that was coming out of her, leftover from her childhood and early adulthood.' All that chemically chlorine gas that has accumulated in your body over the years. It's a frightening thought.

But there is something else we need to check. Are there toxins in the water? Here we encounter a new problem: what do they mean by toxin? I've asked the manufacturers of many detox products this question time and again, but they demur. They wave their hands, they talk about stressful modern lifestyles, they talk about pollution, they talk about junk food, but they will not tell me the name of a single chemical which I can measure. 'What toxins are being extracted from the body with your treatment?' I ask. 'Tell me what is in the water, and I will look for it in a laboratory.' I have never been given an answer.

After much of their hedging and fudging, I chose two chemicals pretty much at random: creatinine and urea. These are common breakdown products from your body's metabolism, and your kidneys get rid of them in urine. Through a friend, I went for a genuine Aqua Detox treatment, took a sample of brown water, and used the disproportionately state-of-the-art analytic facilities of St Mary's Hospital in London to hunt for these two chemical 'toxins'. There were no toxins in the water. Just lots of brown, rusty iron.

Now, with findings like these, scientists might take a step back, and revise their ideas about what is going on with the footbaths. We don't really expect the manufacturers to do that, but what they say in response to these findings is very interesting, at least to me, because it sets up a pattern that we will see repeated throughout the world of pseudoscience: instead of addressing the criticisms, or embracing the new findings in a new model, they seem to shift the goalposts and retreat, crucially, into *untestable positions*.

Some of them now deny that toxins come out in the footbath (which would stop me measuring them): your body is somehow informed that it is time to release toxins in the normal way – whatever that is, and whatever the toxins are – only more so. Some of them now admit that the water goes a bit brown without your feet in it, but 'not as much'. Many of them tell lengthy stories about the 'bioenergetic field', which they say cannot be measured, except by how well you are feeling. All of them talk about how stressful modern life is.

That may well be true. But it has nothing to do with their foot bath, which is all about theatre: and theatre is the common theme for all detox products, as we will see. On with the brown goo.

Ear candles

You might think that Hopi Ear Candles are an easy target. But their efficacy has still been cheerfully promoted by the *Independent*, the *Observer* and the BBC, to name just a few respected news outlets. Since these people are the authoritative purveyors of scientific information, I'll let the BBC explain how these hollow wax tubes will detox your body:

> The candles work by vaporising their ingredients once lit, causing convectional air flow towards the first chamber of the ear. The candle creates a mild suction which lets the vapours gently massage the eardrum and auditory canal. Once the candle is placed in the ear it forms a seal which enables wax and other impurities to be drawn out of the ear.

The proof comes when you open a candle up, and discover that it is filled with a familiar waxy orange substance, which must surely be earwax. If you'd like to test this yourself, you will need: an ear, a clothes peg, some Blu Tack, a dusty floor, some scissors, and two ear candles. I recommend OTOSAN because of their strapline ('The ear is the gateway to the soul').

If you light one ear candle, and hold it over some dust, you will find little evidence of any suction. Before you rush to publish your finding in a peer-reviewed academic journal, someone has beaten you to it: a paper published in the medical journal *Laryngoscope* used expensive tympanometry equipment and found – as you have – that ear candles exert no suction. There is no truth to the claim that doctors dismiss alternative therapies out of hand.

But what if the wax and toxins are being drawn into the candle by some other, more esoteric route, as is often claimed?

For this you will need to do something called a controlled experiment, comparing the results of two different situations, where one is the experimental condition, the other is the 'control' condition, and the only difference is the thing you're interested in testing. This is why you have two candles.

Put one ear candle in someone's ear, as per the manufacturer's instructions, and leave it there until it burns down.* Put the other candle in the clothes peg, and stand it upright using the Blu Tack: this is the 'control arm' in your experiment. The point of a control is simple: we need to minimise the differences between the two set-ups, so that the only real difference between them is the single factor you're studying, which in this case must be: 'Is it my ear that produces the orange goo?'

Take your two candles back inside and cut them open. In the 'ear' candle, you will find a waxy orange substance. In the 'picnic table control', you will find a waxy orange substance. There is only one internationally recognised method for identifying something as earwax: pick some up on the end of your finger, and touch it with your tongue. If your experiment had the same results as mine, both of them taste a lot like candle wax.

Does the ear candle remove earwax from your ears? You can't tell, but a published study followed patients during a full programme of ear candling, and found no reduction. For all that you might have learnt something useful here about the experimental method, there is something more significant you should have picked up: it is expensive, tedious and time-consuming to test every whim concocted out of thin air by therapists selling unlikely miracle cures. But it can be done, and it is done.

---

* Be careful. One paper surveyed 122 ENT doctors, and collected twenty-one cases of serious injury from burning wax falling onto the eardrum during ear-candle treatment.

Detox patches and the 'hassle barrier'

Last in our brown-sludge detox triptych comes the detox foot patch. These are available in most high-street health-food stores, or from your local Avon lady (this is true). They look like teabags, with a foil backing which you stick onto your foot using an adhesive edging before you get into bed. When you wake up the next morning there is a strange-smelling, sticky brown sludge attached to the bottom of your foot, and inside the teabag. This sludge – you may spot a pattern here – is said to be 'toxins'. Except it's not. By now you can probably come up with a quick experiment to show that. I'll give you one option in a footnote.*

An experiment is one way of determining whether an observable effect – sludge – is related to a given process. But you can also pull things apart on a more theoretical level. If you examine the list of ingredients in these patches, you will see that they have been very carefully designed.

The first thing on the list is 'pyroligneous acid', or wood vinegar. This is a brown powder which is highly 'hygroscopic', a word which simply means that it attracts and absorbs water, like those little silicon bags that come in electronic equipment packaging. If there is any moisture around, wood vinegar will absorb it, and make a brown mush which feels warm against your skin.

What is the other major ingredient, impressively listed as 'hydrolysed carbohydrate'? A carbohydrate is a long string of sugar molecules all stuck together. Starch is a carbohydrate, for example, and in your body this is broken down gradually into the individual sugar molecules by your digestive enzymes, so that you can absorb it. The process of breaking down a carbo-

---

* If you take one of these bags and squirt some water onto it, then pop a nice hot cup of tea on top of it and wait for ten minutes, you'll see brown sludge forming. There are no toxins in porcelain.

hydrate molecule into its individual sugars is called 'hydrolysis'. So 'hydrolysed carbohydrate', as you might have worked out by now, for all that it sounds sciencey, basically means 'sugar'. Obviously sugar goes sticky in sweat.

Is there anything more to these patches than that? Yes. There is a new device which we should call 'the hassle barrier', another recurring theme in the more advanced forms of foolishness which we shall be reviewing later. There are huge numbers of different brands, and many of them offer excellent and lengthy documents full of science to prove that they work: they have diagrams and graphs, and the appearance of scienciness; but the key elements are missing. There are experiments, they say, which prove that detox patches do something … but they don't tell you what these experiments consisted of, or what their 'methods' were, they only offer decorous graphs of 'results'.

To focus on the methods is to miss the point of these apparent 'experiments': they aren't about the methods, they're about the positive result, the graph, and the appearance of science. These are superficially plausible totems to frighten off a questioning journalist, a *hassle barrier*, and this is another recurring theme which we will see – in more complex forms – around many of the more advanced areas of bad science. You will come to love the details.

If it's not science, what is it?

Find out if drinking urine, balancing on mountain ledges and genital weightlifting really did change their lives forever.

Channel 4's *Extreme Celebrity Detox*

These are the absurd extremes of detox, but they speak of the larger market, the antioxidant pills, the potions, the books, the juices, the five-day 'programmes', the pipes up the bum and the dreary TV shows, all of which we will torpedo, mostly in a later chapter on nutritionism. But there is something important happening here, with detox, and I don't think it's enough just to say, 'All this is nonsense.'

The detox phenomenon is interesting because it represents one of the most grandiose innovations of marketers, lifestyle gurus, and alternative therapists: the invention of a whole new physiological process. In terms of basic human biochemistry, detox is a meaningless concept. It doesn't cleave nature at the joints. There is nothing on the 'detox system' in a medical textbook. That burgers and beer can have negative effects on your body is certainly true, for a number of reasons; but the notion that they leave a specific residue, which can be extruded by a specific process, a physiological system called detox, is a marketing invention.

If you look at a metabolic flow chart, the gigantic wall-sized maps of all the molecules in your body, detailing the way that food is broken down into its constituent parts, and then those components are converted between each other, and then those new building blocks are assembled into muscle, and bone, and tongue, and bile, and sweat, and bogey, and hair, and skin, and sperm, and brain, and everything that makes you you, it's hard to pick out one thing that is the 'detox system'.

Because it has no scientific meaning, detox is much better understood as a cultural product. Like the best pseudoscientific inventions, it deliberately blends useful common sense with outlandish, medicalised fantasy. In some respects, how much you buy into this reflects how self-dramatising you want to be; or in less damning terms, how much you enjoy ritual in your daily life. When I go through busy periods of partying, drinking, sleep deprivation and convenience eating, I usually decide – eventually – that I need a bit of a rest. So I have a few nights in, reading at home, and eating more salad than usual. Models and celebrities, meanwhile, 'detox'.

On one thing we must be absolutely clear, because this is a

recurring theme throughout the world of bad science. There is nothing wrong with the notion of eating healthily and abstaining from various risk factors for ill health like excessive alcohol use. But that is not what detox is about: these are quick-fix health drives, constructed from the outset as short-term, while lifestyle risk factors for ill health have their impact over a lifetime. But I am even willing to agree that some people might try a five-day detox and remember (or even learn) what it's like to eat vegetables, and that gets no criticism from me.

What's wrong is to pretend that these rituals are based in science, or even that they are new. Almost every religion and culture has some form of purification or abstinence ritual, with fasting, a change in diet, bathing, or any number of other interventions, most of which are dressed up in mumbo jumbo. They're not presented as science, because they come from an era before scientific terms entered the lexicon: but still, Yom Kippur in Judaism, Ramadan in Islam, and all manner of other similar rituals in Christianity, Hinduism, the Baha'i faith, Buddhism, Jainism, are each about abstinence and purification (among other things). Such rituals, like detox regimes, are conspicuously and – to some believers too, I'm sure – spuriously precise. Hindu fasts, for example, if strictly observed, run from the previous day's sunset until *forty-eight minutes* after the next day's sunrise.

Purification and redemption are such recurrent themes in ritual because there is a clear and ubiquitous need for them: we all do regrettable things as a result of our own circumstances, and new rituals are frequently invented in response to new circumstances. In Angola and Mozambique, purification and cleansing rituals have arisen for children affected by war, particularly former child soldiers. These are healing rituals, where the child is purged and purified of sin and guilt, of the 'contamination' of war and death (contamination is a recurring metaphor in all cultures, for obvious reasons); the child is also protected from the consequences of his previous actions, which is to say, he is protected from retaliation by the avenging spirits of those he has killed. As a World Bank report put it in 1999:

> These cleansing and purification rituals for child soldiers have the appearance of what anthropologists call rites of transition. That is, the child undergoes a symbolic change of status from someone who has existed in a realm of sanctioned norm-violation or norm-suspension (i.e. killing, war) to someone who must now live in a realm of peaceful behavioural and social norms, and conform to these.

I don't think I'm stretching this too far. In what we call the developed Western world, we seek redemption and purification from the more extreme forms of our material indulgence: we fill our faces with drugs, drink, bad food and other indulgences, we know it's wrong, and we crave ritualistic protection from the consequences, a public 'transitional ritual' commemorating our return to healthier behavioural norms.

The presentation of these purification diets and rituals has always been a product of their time and place, and now that science is our dominant explanatory framework for the natural and moral world, for right or wrong, it's natural that we should bolt a bastardised pseudoscientific justification onto our redemption. Like so much of the nonsense in bad science, 'detox' pseudoscience isn't something done *to* us, by venal and exploitative outsiders: it is a cultural product, a recurring theme, and we do it to ourselves.

2

Brain Gym

Under normal circumstances this should be the part of the book where I fall into a rage over creationism, to gales of applause, even though it's a marginal issue in British schools. But if you want an example from closer to home, there is a vast empire of pseudoscience being peddled, for hard cash, in state schools up and down the country. It's called Brain Gym, it is pervasive throughout the state education system, it's swallowed whole by teachers, it's presented directly to the children they teach, and it's riddled with transparent, shameful and embarrassing nonsense.

At the heart of Brain Gym is a string of complicated and proprietary exercises for kids which 'enhance the experience of whole brain learning'. They're very keen on water, for example. 'Drink a glass of water before Brain Gym activities', they say. 'As it is a major component of blood, water is vital for transporting oxygen to the brain.' Heaven forbid that your blood should dry out. This water should be held in your mouth, they say, because then it can be absorbed *directly* from there into your brain.

Is there anything else you can do to get blood and oxygen to your brain more efficiently? Yes, an exercise called 'Brain Buttons': 'Make a "C" shape with your thumb and forefinger and place on either side of the breastbone just below the collarbone.

Gently rub for twenty or thirty seconds whilst placing your other hand over your navel. Change hands and repeat. This exercise stimulates the flow of oxygen carrying blood through the carotid arteries to the brain to awaken it and increase concentration and relaxation.' Why? 'Brain buttons lie directly over and stimulate the carotid arteries.'

Children can be disgusting, and often they can develop extraordinary talents, but I'm yet to meet any child who can stimulate his carotid arteries inside his ribcage. That's probably going to need the sharp scissors that only mummy can use.

You might imagine that this nonsense is a marginal, peripheral trend which I have contrived to find in a small number of isolated, misguided schools. But no. Brain Gym is practised in hundreds if not thousands of mainstream state schools throughout the country. As of today I have a list of over four hundred schools which mention it specifically by name on their websites, and many, many others will also be using it. Ask if they do it at your school. I'd be genuinely interested to know their reaction.

Brain Gym is promoted by local education authorities, funded by the government, and the training counts as continuing professional development for teachers. But it doesn't end locally. You will find Brain Gym being promoted on the Department for Education and Skills website, in all kinds of different places, and it pops up repeatedly as a tool for promoting 'inclusivity', as if pushing pseudoscience at children is somehow going to ameliorate social inequality, rather than worsen it. This is a vast empire of nonsense infecting the entirety of the British education system, from the smallest primary school to central government, and nobody seems to notice or care.

Perhaps if they could just do the 'hook-up' exercises on page 31 of the *Brain Gym Teacher's Manual* (where you press your fingers against each other in odd contorted patterns) this would 'connect the electrical circuits in the body, containing and thus focusing both attention and disorganised energy', and they would finally see sense. Perhaps if they wiggled their ears with their fingers as per the Brain Gym textbook it would 'stimulate the reticular formation of the brain to tune out distracting, irrelevant sounds and tune into language'.

The same teacher who explains to your children how blood is pumped around the lungs and then the body by the heart is also telling them that when they do the 'Energizer' exercise (which is far too complicated to describe), 'this back and forward movement of the head increases the circulation to the frontal lobe for greater comprehension and rational thinking'. Most frighteningly, this teacher sat through a class, being taught this nonsense by a Brain Gym instructor, without challenging or questioning it.

In some respects the issues here are similar to those in the chapter on detox: if you just want to do a breathing exercise, then that's great. But the creators of Brain Gym go much further. Their special, proprietary, theatrical yawn will lead to 'increased oxidation for efficient relaxed functioning'. Oxidation is what causes rusting. It is not the same as oxygenation, which I suppose is what they mean. (And even if they are talking about oxygenation, you don't need to do a funny yawn to get oxygen into your blood: like most other wild animals, children have a perfectly adequate and fascinating physiological system in place to regulate their blood oxygen and carbon dioxide levels, and I'm sure many of them would rather be taught about that, and indeed about the role of electricity in the body, or any of the other things Brain Gym confusedly jumbles up, than this transparent pseudoscientific nonsense.)

How can this nonsense be so widespread in schools? One obvious explanation is that the teachers have been blinded by all these clever long phrases like 'reticular formation' and 'increased oxidation'. As it happens, this very phenomenon has been studied in a fascinating set of experiments from the March

2008 edition of the *Journal of Cognitive Neuroscience*, which elegantly demonstrated that people will buy into bogus explanations much more readily when they are dressed up with a few technical words from the world of neuroscience.

Subjects were given descriptions of various phenomena from the world of psychology, and then randomly offered one of four explanations for them. The explanations either contained neuroscience or didn't, and were either 'good' explanations or 'bad' ones (bad ones being, for example, simply circular restatements of the phenomenon itself, or empty words).

Here is one of the scenarios. Experiments have shown that people are quite bad at estimating the knowledge of others: if *we* know the answer to a question about a piece of trivia, we overestimate the extent to which other people will know that answer too. In the experiment a 'without neuroscience' explanation for this phenomenon was: 'The researchers claim that this [overestimation] happens because subjects have trouble switching their point of view to consider what someone else might know, mistakenly projecting their own knowledge onto others.' (This was a 'good' explanation.)

A 'with neuroscience' explanation – and a cruddy one too – was this: 'Brain scans indicate that this [overestimation] happens because of the frontal lobe brain circuitry known to be involved in self-knowledge. Subjects make more mistakes when they have to judge the knowledge of others. People are much better at judging what they themselves know.' Very little is added by this explanation, as you can see. Furthermore, the neuroscience information is merely decorative, and irrelevant to the explanation's logic.

The subjects in the experiment were from three groups: everyday people, neuroscience students, and neuroscience academics, and they performed very differently. All three groups judged good explanations as more satisfying than bad ones, but the subjects in the two non-expert groups judged that the explanations *with* the logically irrelevant neurosciencey information were more satisfying than the explanations *without* the spurious neuroscience. What's more, the spurious neuroscience had a particularly strong effect on people's judgements of 'bad' explanations. Quacks, of course, are well aware of this, and have been adding sciencey-sounding explanations to their products for as long as quackery has existed, as a means to bolster their authority over the patient (in an era, interestingly, when doctors have struggled to inform patients more, and to engage them in decisions about their own treatment).

It's interesting to think about why this kind of decoration is so seductive, and to people who should know better. Firstly, the very presence of neuroscience information might be seen as a surrogate marker of a 'good' explanation, regardless of what is actually said. As the researchers say, 'Something about seeing neuroscience information may encourage people to believe they have received a scientific explanation when they have not.'

But more clues can be found in the extensive literature on irrationality. People tend, for example, to rate longer explanations as being more similar to 'experts' explanations'. There is also the 'seductive details' effect: if you present related (but logically irrelevant) details to people as part of an argument, this seems to make it more difficult for them to encode, and later recall, the main argument of a text, because their attention is diverted.

More than this, perhaps we all have a rather Victorian fetish for reductionist explanations about the world. They just feel neat, somehow. When we read the neurosciencey language in the 'bogus neuroscience explanations' experiment – and in the Brain Gym literature – we feel as if we have been given a physical explanation for a behavioural phenomenon ('an exercise break in class is refreshing'). We have somehow made behavioural phenomena feel connected to a larger explanatory system, the physical sciences, a world of certainty, graphs and unambiguous data. It feels like progress. In fact, as is often the case with spurious certainty, it's the very opposite.

Again, we should focus for a moment on what is good about Brain Gym, because when you strip away the nonsense, it advocates regular breaks, intermittent light exercise, and drinking plenty of water. This is all entirely sensible.

But Brain Gym perfectly illustrates two more recurring themes from the industry of pseudoscience. The first is this: you can use hocus pocus – or what Plato euphemistically called a 'noble myth' – to make people do something fairly sensible like drink some water and have an exercise break. You will have your own view on when this is justified and proportionate (perhaps factoring in issues like whether it's necessary, and the side-effects of pandering to nonsense), but it strikes me that in the case of Brain Gym, this is not a close call: children are predisposed to learn about the world from adults, and specifically from teachers; they are sponges for information, for ways of seeing, and authority figures who fill their heads with nonsense are sowing the ground, I would say, for a lifetime of exploitation.

The second theme is perhaps more interesting: the proprietorialisation of common sense. You can take a perfectly sensible intervention, like a glass of water and an exercise break, but add nonsense, make it sound more technical, and make yourself sound clever. This will enhance the placebo effect, but you might also wonder whether the primary goal is something much more cynical and lucrative: to make common sense copyrightable, unique, patented, and *owned*.

We will see this time and again, on a grander scale, in the work of dubious healthcare practitioners, and specifically in the field of 'nutritionism', because scientific knowledge – and sensible dietary advice – is free and in the public domain. Anyone can use it, understand it, sell it, or simply give it away. Most people know what constitutes a healthy diet already. If you want to make money out of it, you have to make a space for yourself in the market: and to do this, you must overcomplicate it, attach your own dubious stamp.

Is there any harm in this process? Well, it's certainly wasteful, and even in the decadent West, as we enter a probable recession, it does seem peculiar to give money away for basic diet advice, or exercise breaks at school. But there are other hidden dangers, which are far more corrosive. This process of professionalising the obvious fosters a sense of mystery around science, and health advice, which is unnecessary and destructive. More than anything, more than the unnecessary ownership of the obvious, it is disempowering. All too often this spurious privatisation of common sense is happening in areas where we could be taking control, doing it ourselves, feeling our own potency and our ability to make sensible decisions; instead we are fostering our dependence on expensive outside systems and people.

But what's most frightening is the way that pseudoscience makes your head go soggy. Debunking Brain Gym, let me remind you, does not require high-end, specialist knowledge. We are talking about a programme which claims that 'processed foods do not contain water', possibly the single most rapidly falsifiable statement I've seen all week. What about soup? 'All other liquids are processed in the body as food, and do not serve the body's water needs.'

This is an organisation at the edges of reason, but it is operating in countless British schools. When I wrote about Brain Gym in my newspaper column in 2005, saying 'exercise breaks good, pseudoscientific nonsense laughable', while many teachers erupted with delight, many were outraged and 'disgusted' by what they decided was an attack on exercises which they experienced as helpful. One – an assistant head teacher, no less – demanded: 'From what I can gather you have visited no classrooms, interviewed no teachers nor questioned any children, let alone had a conversation with any of a number of specialists in this field?'

Do I need to visit a classroom to find out if there is water in processed food? No. If I meet a 'specialist' who tells me that a child can massage both carotid arteries through the ribcage (without scissors), what will I say to them? If I meet a teacher who thinks that touching your fingers together will connect the electrical circuit of the body, where do we go from there?

I'd like to imagine that we live in a country where teachers might have the nous to spot this nonsense and stop it in its tracks. If I was a different kind of person I'd be angrily confronting the responsible government departments, and demanding to know what they were going to do about it, and reporting back to you with their mumbling and shamed defence. But I am not that kind of journalist, and Brain Gym is so obviously, transparently foolish that nothing they could say could possibly justify the claims made on its behalf. Just one thing gives me hope, and that is the steady trickle of emails I receive on the subject from children, ecstatic with delight at the stupidity of their teachers:

> I'd like to submit to Bad Science my teacher who gave us a handout which says that 'Water is best absorbed by the body when provided in frequent small amounts.' What I want to know is this. If I drink too much in one go, will it leak out of my arsehole instead?
>
> *'Anton', 2006*

Thank you Anton.

3

The Progenium XY Complex

I have great respect for the manufacturers of cosmetics. They are at the other end of the spectrum from the detox industry: this is a tightly regulated industry, with big money to be made from nonsense, and so we find large, well-organised teams from international biotech firms generating elegant, distracting, suggestive, but utterly defensible pseudoscience. After the childishness of Brain Gym, we can now raise our game.

Before we start, it's important to understand how cosmetics – and specifically moisturising creams – actually work, because there should be no mystery here. Firstly, you want your expensive cream to hydrate your skin. They all do that, and Vaseline does the job very well: in fact, much of the important early cosmetics research was about preserving the moisturising properties of Vaseline, while avoiding its greasiness, and this technical mountain was scaled several decades ago. Hydrobase, at around £10 for a half-litre tub from your chemist, will do the job excellently.

If you really want to, you can replicate this by making your own moisturiser at home: you're aiming for a mix of water and oil, but one that's 'emulsified', which is to say, nicely mixed up. When I was involved in hippy street theatre – and I'm being entirely serious here – we made moisturiser from equal parts of olive oil, coconut oil, honey and rosewater (tap water is fine too). Beeswax is better than honey as an emulsifier, and you can modify the cream's consistency for yourself: more beeswax will make it firmer, more oil will make it softer, and more water makes it sort of fluffier, but increases the risk of the ingredients separating out. Get all your ingredients lightly heated, but separately, stir the oil into the wax, beating all the time, and then stir in the water. Stick it in a jar, and keep for three months in the fridge.

The creams in your local pharmacy seem to go way beyond this. They are filled with magic ingredients: Regenium XY technology, Nutrileum complex, RoC Retinol Correxion, Vita-Niacin, Covabeads, ATP Stimuline and Tenseur Peptidique Végétal. Surely you could never replicate that in your kitchen, or with creams that cost as much by the gallon as these ones cost for a squirt of the tiny tube? What are these magic ingredients? And what do they do?

There are basically three groups of ingredients in moisturising cream. Firstly there are powerful chemicals, like alpha-hydroxy acids, high levels of vitamin C, or molecular variations on the theme of vitamin A. These have genuinely been shown to make your skin seem more youthful, but they are only effective at such high concentrations, or high acidity levels, that the creams cause irritation, stinging, burning and redness. They were the great white hope in the 1990s, but now they've all had to be massively watered down by law, unless on prescription. No free lunch, and no effects without side-effects, as usual.

Companies still name them on the label, wallowing in the glory of their efficacy at higher potencies, because you don't have to give the doses of your ingredients, only their ranked order. But these chemicals are usually in your cream at talismanic concentrations, for show only. The claims made on the various bottles and tubes are from the halcyon days of effective and high-potency acidic creams, but that's hard to tell, because they are usually based on privately funded and published studies, done by the industry, and rarely available in their complete published forms, as a proper academic paper should be, so that you can check the working. Of course, forgetting that technical stuff, most of the 'evidence' quoted in cream adverts is from subjective reports, where 'seven out of ten people who received free pots of cream were very pleased with the results'.

The second ingredient in almost all posh creams is one which does kind of work: cooked and mashed-up vegetable protein (hydrolysed X-microprotein nutricomplexes, Tenseur Peptidique Végétal, or whatever they're calling them this month). These are long, soggy chains of amino acids, which swim around in the cream, languorously stretched out in the moistness of it all. When the cream dries on your face, these long, soggy chains contract and tighten: the slightly unpleasant taut sensation you get on your face when you wear these creams is from the protein chains contracting all over your skin, which temporarily shrinks your finer wrinkles. It is a fleeting but immediate pay-off from using the expensive creams, but it wouldn't help you choose between them, since almost all of them contain mashed-up protein chains.

Finally there is the huge list of esoteric ingredients, tossed in on a prayer, with suggestive language elegantly woven around them in a way that allows you to believe that all kinds of claims are being made.

Classically, cosmetics companies will take highly theoretical, textbookish information about the way that cells work – the components at a molecular level, or the behaviour of cells in a glass dish – and then pretend it's the same as the ultimate issue of whether something makes you look nice. 'This molecular component,' they say, with a flourish, 'is crucial for collagen formation.' And that will be perfectly true (along with many other amino acids which are used by your body to assemble protein in joints, skin, and everywhere else), but there is no reason to believe that anyone is deficient in it, or that smearing it on your face will make any difference to your appearance. In general, you don't absorb things very well through your skin, because its purpose is to be relatively impermeable. When you sit in a bath of baked beans for charity you do not get fat, nor do you start farting.

Despite this, on any trip to the chemist (I recommend it) you can find a phenomenal array of magic ingredients on the market. Valmont Cellular DNA Complex is made from 'specially treated salmon roe DNA' ('Unfortunately, smearing salmon on your face won't have quite the same effect,' said *The Times* in their review), but it's spectacularly unlikely that DNA – a very large molecule indeed – would be absorbed by your skin, or indeed be any use for the synthetic activity happening in it, even if it was. You're probably not short of the building blocks of DNA in your body. There's a hell of a lot of it in there already.

Thinking it through, if salmon DNA *was* absorbed whole by your skin, then you would be absorbing alien, or rather fish, design blueprints into your cells – that is, the instructions for making fish cells, which might not be helpful for you as a human. It would also be a surprise if the DNA was digested into its constituent elements in your skin (your gut, though, is specifically adapted for digesting large molecules, using digestive enzymes which break them up into their constituent parts before absorption).

The simple theme running through all these products is that you can hoodwink your body, when in reality there are finely tuned 'homeostatic' mechanisms, huge, elaborate systems with feedback and measuring devices, constantly calibrating and recalibrating the amounts of various different chemical constituents being sent to different parts of your body. If anything, interfering with that system is likely to have the opposite of the simplistic effects claimed.

As the perfect example, there are huge numbers of creams (and other beauty treatments) claiming to deliver oxygen directly to your skin. Many of the creams contain peroxide, which, if you really want to persuade yourself of its efficacy, has a chemical formula of H_2O_2, and could fancifully be conceived of as water 'with some extra oxygen', although chemical formulae don't really work that way – after all, a pile of rust is an iron bridge 'with some extra oxygen', and you wouldn't imagine it would oxygenate your skin.

Even if we give them the benefit of the doubt and pretend that these treatments really will deliver oxygen to the surface of the skin, and that this will penetrate meaningfully into the cells, what good would that do? Your body is constantly monitoring the amount of blood and nutrients it's supplying to tissues, and the quantity of tiny capillary arteries feeding a given area, and more vessels will grow towards areas with low oxygen, because that is a good index of whether more blood supply is needed. Even if the claim about oxygen in cream penetrating your tissues were true, your body would simply downregulate the supply of blood to that part of skin, scoring a homeostatic own goal. In reality, hydrogen peroxide is simply a corrosive chemical that gives you a light chemical burn at low strengths. This might explain that fresh, glowing feeling.

These details generalise to most of the claims made on packaging. Look closely at the label or advert, and you will routinely find that you are being played in an elaborate semantic game, with the complicity of the regulators: it's rare to find an explicit claim, that rubbing this particular magic ingredient on your face will make you look better. The claim is made for the cream *as a whole*, and it is true for the cream as a whole, because as you now know, all moisturising creams – even a cheap litre tub of Diprobase – will moisturise.

Once you know this, shopping becomes marginally more interesting. The link between the magic ingredient and efficacy is made only in the customer's mind, and reading through the manufacturer's claims you can see that they have been carefully reviewed by a small army of consultants to ensure that the label is highly suggestive, but also – to the eye of an informed pedant – semantically and legally watertight. (If you want to make a living in this field, I would recommend the well-trodden career path: a spell in trading standards, advertising standards, or any other regulatory body, before going on to work as a consultant to industry.)

So what's wrong with this kind of spin? We should be clear on one thing: I'm not on a consumer crusade. Just like the National Lottery, the cosmetics industry is playing on people's dreams, and people are free to waste their money. I can very happily view posh cosmetics – and other forms of quackery – as a special, self-administered, voluntary tax on people who don't understand science properly. I would also be the first to agree that people don't buy expensive cosmetics simply because they have a belief in their efficacy, because it's 'a bit more complicated than that': these are luxury goods, status items, and they are bought for all kinds of interesting reasons.

But it's not entirely morally neutral. Firstly, the manufacturers of these products sell shortcuts to smokers and the obese; they sell the idea that a healthy body can be attained by using expensive potions, rather than simple old-fashioned exercise and eating your greens. This is a recurring theme throughout the world of bad science.

More than that, these adverts sell a dubious world view. They sell the idea that science is not about the delicate relationship between evidence and theory. They suggest, instead, with all the might of their international advertising budgets, their Microcellular Complexes, their Neutrillium XY, their Tenseur Peptidique Végétal and the rest, that science is about impenetrable nonsense involving equations, molecules, sciencey diagrams, sweeping didactic statements from authority figures in white coats, and that this sciencey-sounding stuff might just as well be made up, concocted, confabulated out of thin air, in order to make money. They sell the idea that science is incomprehensible, with all their might, and they sell this idea mainly to attractive young women, who are disappointingly under-represented in the sciences.

In fact, they sell the world view of 'Teen Talk Barbie' from Mattel, who shipped with a sweet little voice circuit inside her so she could say things like, 'Math class is tough!', 'I love shopping!' and 'Will we ever have enough clothes?' when you pressed her buttons. In December 1992 the feminist direct-action Barbie Liberation Organization switched the voice circuits of hundreds of Teen Talk Barbies and GI Joe dolls in American shops. On Christmas Day Barbie said 'Dead men tell no lies' in a nice assertive voice, and the boys got soldiers under the tree telling them 'Math class is tough!' and asking 'Wanna go shopping?'

The work of the BLO is not yet done.

4

Homeopathy

And now for the meat. But before we take a single step into this arena, we should be clear on one thing: despite what you might think, I'm not desperately interested in Complementary and Alternative Medicine (a dubious piece of phraseological rebranding in itself). I am interested in the role of medicine, our beliefs about the body and healing, and I am fascinated – in my day job – by the intricacies of how we can gather evidence for the benefits and risks of a given intervention.

Homeopathy, in all of this, is simply our tool.

So here we address one of the most important issues in science: how do we know if an intervention works? Whether it's a face cream, a detox regime, a school exercise, a vitamin pill, a parenting programme or a heart-attack drug, the skills involved in testing an intervention are all the same. Homeopathy makes the clearest teaching device for evidence-based medicine for one simple reason: homeopaths give out little sugar pills, and pills are the easiest thing in the world to study.

By the end of this section you will know more about evidence-based medicine and trial design than the average doctor. You will understand how trials can go wrong, and give false positive results, how the placebo effect works, and why we tend to overestimate the efficacy of pills. More importantly, you will also see how a health myth can be created, fostered and maintained by the alternative medicine industry, using all the same tricks on you, the public, which big pharma uses on doctors. This is about something much bigger than homeopathy.

What is homeopathy?

Homeopathy is perhaps the paradigmatic example of an alternative therapy: it claims the authority of a rich historical heritage, but its history is routinely rewritten for the PR needs of a contemporary market; it has an elaborate and sciencey-sounding framework for how it works, without scientific evidence to demonstrate its veracity; and its proponents are quite clear that the pills will make you better, when in fact they have been thoroughly researched, with innumerable trials, and have been found to perform no better than placebo.

Homeopathy was devised by a German doctor named Samuel Hahnemann in the late eighteenth century. At a time when mainstream medicine consisted of blood-letting, purging and various other ineffective and dangerous evils, when new treatments were conjured up out of thin air by arbitrary authority figures who called themselves 'doctors', often with little evidence to support them, homeopathy would have seemed fairly reasonable.

Hahnemann's theories differed from the competition because he decided – and there's no better word for it – that if he could find a substance which would induce the symptoms of a disease in a healthy individual, then it could be used to treat the same symptoms in a sick person. His first homeopathic remedy was Cinchona bark, which was suggested as a treatment for malaria. He took some himself, at a high dose, and experienced symptoms which he decided were similar to those of malaria itself:

My feet and finger-tips at once became cold; I grew languid and
drowsy; my heart began to palpitate; my pulse became hard and
quick; an intolerable anxiety and trembling arose ... prostration
... pulsation in the head, redness in the cheek and raging thirst
... intermittent fever ... stupefaction ... rigidity ...

and so on.

Hahnemann assumed that everyone would experience these
symptoms if they took Cinchona (although there's some
evidence that he just experienced an idiosyncratic adverse reaction). More importantly, he also decided that if he gave a tiny
amount of Cinchona to someone with malaria, it would treat,
rather than cause, the malaria symptoms. The theory of 'like
cures like' which he conjured up on that day is, in essence, the
first principle of homeopathy.*

Giving out chemicals and herbs could be a dangerous business, since they can have genuine effects on the body (they
induce symptoms, as Hahnemann identified). But he solved
that problem with his second great inspiration, and the key
feature of homeopathy that most people would recognise
today: he decided – again, that's the only word for it – that if you
diluted a substance, this would 'potentise' its ability to cure
symptoms, 'enhancing' its 'spirit-like medicinal powers', and at
the same time, as luck would have it, also reducing its side-
effects. In fact he went further than this: the more you dilute a
substance, the more powerful it becomes at treating the symptoms it would otherwise induce.

Simple dilutions were not enough. Hahnemann decided that
the process had to be performed in a very specific way, with an
eye on brand identity, or a sense of ritual and occasion, so he
devised a process called 'succussion'. With each dilution the glass

---

* At proper high doses, Cinchona contains quinine, which can genuinely be
used to treat malaria, although most malarial parasites are immune to it now.

vessel containing the remedy is shaken by ten firm strikes against 'a hard but elastic object'. For this purpose Hahnemann had a saddlemaker construct a bespoke wooden striking board, covered in leather on one side, and stuffed with horsehair. These ten firm strikes are still carried out in homeopathy pill factories today, sometimes by elaborate, specially constructed robots.

Homeopaths have developed a wide range of remedies over the years, and the process of developing them has come to be called, rather grandly, 'proving' (from the German *Prufung*). A group of volunteers, anywhere from one person to a couple of dozen, come together and take six doses of the remedy being 'proved', at a range of dilutions, over the course of two days, keeping a diary of the mental, physical and emotional sensations, including dreams, experienced over this time. At the end of the proving, the 'master prover' will collate the information from the diaries, and this long, unsystematic list of symptoms and dreams from a small number of people will become the 'symptom picture' for that remedy, written in a big book and revered, in some cases, for all time. When you go to a homeopath, he or she will try to match your symptoms to the ones caused by a remedy in a proving.

There are obvious problems with this system. For a start, you can't be sure if the experiences the 'provers' are having are caused by the substance they're taking, or by something entirely unrelated. It might be a 'nocebo' effect, the opposite of placebo, where people feel bad because they're expecting to (I bet I could make you feel nauseous right now by telling you some home truths about how your last processed meal was made); it might be a form of group hysteria ('Are there fleas in this sofa?'); one of them might experience a tummy ache that was coming on anyway; or they might all get the same mild cold together; and so on.

But homeopaths have been very successful at marketing these 'provings' as valid scientific investigations. If you go to

Boots the Chemist's website, www.bootslearningstore.co.uk, for example, and take their 16-plus teaching module for children on alternative therapies, you will see, amongst the other gobbledegook about homeopathic remedies, that they are teaching how Hahnemann's provings were 'clinical trials'. This is not true, as you can now see, and that is not uncommon.

Hahnemann professed, and indeed recommended, complete ignorance of the physiological processes going on inside the body: he treated it as a black box, with medicines going in and effects coming out, and championed only empirical data, the effects of the medicine on symptoms ('The totality of symptoms and circumstances observed in each individual case,' he said, 'is the one and only indication that can lead us to the choice of the remedy').

This is the polar opposite of the 'Medicine only treats the symptoms, we treat and understand the underlying cause' rhetoric of modern alternative therapists. It's also interesting to note, in these times of 'natural is good', that Hahnemann said nothing about homeopathy being 'natural', and promoted himself as a man of science.

Conventional medicine in Hahnemann's time was obsessed with theory, and was hugely proud of basing its practice on a 'rational' understanding of anatomy and the workings of the body. Medical doctors in the eighteenth century sneeringly accused homeopaths of 'mere empiricism', an over-reliance on observations of people getting better. Now the tables are turned: today the medical profession is frequently happy to accept ignorance of the details of mechanism, as long as trial data shows that treatments are effective (we aim to abandon the ones that aren't), whereas homeopaths rely exclusively on their exotic theories, and ignore the gigantic swathe of negative empirical evidence on their efficacy. It's a small point, perhaps, but these subtle shifts in rhetoric and meaning can be revealing.

The dilution problem

Before we go any further into homeopathy, and look at whether it actually works or not, there is one central problem we need to get out of the way.

Most people know that homeopathic remedies are diluted to such an extent that there will be no molecules of it left in the dose you get. What you might not know is just how far these remedies are diluted. The typical homeopathic dilution is 30C: this means that the original substance has been diluted by one drop in a hundred, thirty times over. In the 'What is homeopathy?' section on the Society of Homeopaths' website, the single largest organisation for homeopaths in the UK will tell you that '30C contains less than one part per million of the original substance.'

'Less than one part per million' is, I would say, something of an understatement: a 30C homeopathic preparation is a dilution of one in 100^{30}, or rather 10^{60}, or one followed by sixty zeroes. To avoid any misunderstandings, this is a dilution of one in 1,000,000,000,000,000,000,000,000,000,000,000,000,000, 000,000,000,000,000,000, or, to phrase it in the Society of Homeopaths' terms, 'one part per million million million million million million million million million million'. This is definitely 'less than one part per million of the original substance'.

For perspective, there are only around 100,000,000,000,000, 000,000,000,000,000,000 molecules of water in an Olympic-sized swimming pool. Imagine a sphere of water with a diameter of 150 million kilometres (the distance from the earth to the sun). It takes light eight minutes to travel that distance. Picture a sphere of water that size, with one molecule of a substance in it: that's a 30C dilution.*

At a homeopathic dilution of 200C (you can buy much higher dilutions from any homeopathic supplier) the treating

---

* For pedants, it's a 30.89C dilution.

substance is diluted more than the total number of atoms in the universe, and by an enormously huge margin. To look at it another way, the universe contains about 3×10^{80} cubic metres of storage space (ideal for starting a family): if it was filled with water, and one molecule of active ingredient, this would make for a rather paltry 55C dilution.

We should remember, though, that the improbability of homeopaths' claims for *how* their pills might work remains fairly inconsequential, and is not central to our main observation, which is that they work no better than placebo. We do not know *how* general anaesthetics work, but we know that they *do* work, and we use them despite our ignorance of the mechanism. I myself have cut deep into a man's abdomen and rummaged around his intestines in an operating theatre – heavily supervised, I hasten to add – while he was knocked out by anaesthetics, and the gaps in our knowledge regarding their mode of action didn't bother either me or the patient at the time.

Moreover, at the time that homeopathy was first devised by Hahnemann, nobody even knew that these problems existed, because the Italian physicist Amadeo Avogadro and his successors hadn't yet worked out how many molecules there are in a given amount of a given substance, let alone how many atoms there are in the universe. We didn't even really know what atoms were.

How have homeopaths dealt with the arrival of this new knowledge? By saying that the absent molecules are irrelevant, because 'water has a memory'. This sounds feasible if you think of a bath, or a test tube full of water. But if you think, at the most basic level, about the scale of these objects, a tiny water molecule isn't going to be deformed by an enormous arnica molecule, and be left with a 'suggestive dent', which is how many homeopaths seem to picture the process. A pea-sized lump of putty cannot take an impression of the surface of your sofa.

Physicists have studied the structure of water very intensively for many decades, and while it is true that water molecules will form structures round a molecule dissolved in them at room temperature, the everyday random motion of water molecules means that these structures are very short-lived, with lifetimes measured in picoseconds, or even less. This is a very restrictive shelf life.

Homeopaths will sometimes pull out anomalous results from physics experiments and suggest that these prove the efficacy of homeopathy. They have fascinating flaws which can be read about elsewhere (frequently the homeopathic substance – which is found on hugely sensitive lab tests to be subtly different from a non-homeopathic dilution – has been prepared in a completely different way, from different stock ingredients, which is then detected by exquisitely sensitive lab equipment). As a ready shorthand, it's also worth noting that the American magician and 'debunker' James Randi has offered a $1 million prize to anyone demonstrating 'anomalous claims' under laboratory conditions, and has specifically stated that anyone could win it by reliably distinguishing a homeopathic preparation from a non-homeopathic one using any method they wish. This $1 million bounty remains unclaimed.

Even if taken at face value, the 'memory of water' claim has large conceptual holes, and most of them you can work out for yourself. If water has a memory, as homeopaths claim, and a one in 10^{60} dilution is fine, then by now all water must surely be a health-giving homeopathic dilution of all the molecules in the world. Water has been sloshing around the globe for a very long time, after all, and the water in my very body as I sit here typing away in London has already been through plenty of other people's bodies before mine. Maybe some of the water molecules sitting in my fingers as I type this sentence are currently in your eyeball. Maybe some of the water molecules fleshing out my neurons as I decide whether to write 'wee' or 'urine' in this sentence are now in the Queen's bladder (God bless her): water is a great leveller, it gets about. Just look at clouds.

How does a water molecule know to forget every other molecule it's seen before? How does it know to treat my bruise with its memory of arnica, rather than a memory of Isaac Asimov's faeces? I wrote this in the newspaper once, and a homeopath complained to the Press Complaints Commission. It's not about the dilution, he said: it's the succussion. You have to bang the flask of water briskly ten times on a leather and horsehair surface, and that's what makes the water remember a molecule. Because I did not mention this, he explained, *I had deliberately made homeopaths sound stupid.* This is another universe of foolishness.

And for all homeopaths' talk about the 'memory of water', we should remember that what you actually take, in general, is a little sugar pill, not a teaspoon of homeopathically diluted water – so they should start thinking about the memory of sugar, too. The memory of sugar, which is remembering something that was being remembered by water (after a dilution greater than the number of atoms in the universe) but then got passed on to the sugar as it dried. I'm trying to be clear, because I don't want any more complaints.

Once this sugar which has remembered something the water was remembering gets into your body, it must have some kind of effect. What would that be? Nobody knows, but you need to take the pills regularly, apparently, in a dosing regime which is suspiciously similar to that for medical drugs (which are given at intervals spaced according to how fast they are broken down and excreted by your body).

I demand a fair trial

These theoretical improbabilities are interesting, but they're not going to win you any arguments: Sir John Forbes, physician to Queen Victoria, pointed out the dilution problem in the nineteenth century, and 150 years later the discussion has not moved on. The real question with homeopathy is very simple: does it work? In fact, how do we know if *any* given treatment is working?

Symptoms are a very subjective thing, so almost every conceivable way of establishing the benefits of any treatment must start with the individual and his or her experience, building from there. Let's imagine we're talking – maybe even arguing – with someone who thinks that homeopathy works, someone who feels it is a positive experience, and who feels they get better, quicker, with homeopathy. They would say: 'All I know is, I feel as if it works. I get better when I take homeopathy.' It seems obvious to them, and to an extent it is. This statement's power, and its flaws, lie in its simplicity. Whatever happens, the statement stands as true.

But you could pop up and say: 'Well, perhaps that was the placebo effect.' Because the placebo effect is far more complex and interesting than most people suspect, going way beyond a mere sugar pill: it's about the whole cultural experience of a treatment, your expectations beforehand, the consultation process you go through while receiving the treatment, and much more.

We know that two sugar pills are a more effective treatment than one sugar pill, for example, and we know that salt-water injections are a more effective treatment for pain than sugar pills, not because salt-water injections have any biological action on the body, but because an injection feels like a more dramatic intervention. We know that the colour of pills, their packaging, how much you pay for them and even the beliefs of the people handing the pills over are all important factors. We know that placebo operations can be effective for knee pain, and even for angina. The placebo effect works on animals and children. It is highly potent, and very sneaky, and you won't know the half of it until you read the 'placebo' chapter in this book.

So when our homeopathy fan says that homeopathic treatment makes them feel better, we might reply: 'I accept that, but perhaps your improvement is because of the placebo effect,' and they cannot answer 'No,' because they have *no possible way of knowing* whether they got better through the placebo effect or not. They cannot tell. The most they can do is restate, in response to your query, their original statement: 'All I know is, I feel as if it works. I get better when I take homeopathy.'

Next, you might say: 'OK, I accept that, but perhaps, also, you feel you're getting better because of "regression to the mean".' This is just one of the many 'cognitive illusions' described in this book, the basic flaws in our reasoning apparatus which lead us to see patterns and connections in the world around us, when closer inspection reveals that in fact there are none.

'Regression to the mean' is basically another phrase for the phenomenon whereby, as alternative therapists like to say, all things have a natural cycle. Let's say you have back pain. It comes and goes. You have good days and bad days, good weeks and bad weeks. When it's at its very worst, it's going to get better, because that's the way things are with your back pain.

Similarly, many illnesses have what is called a 'natural history': they are bad, and then they get better. As Voltaire said: 'The art of medicine consists in amusing the patient while nature cures the disease.' Let's say you have a cold. It's going to get better after a few days, but at the moment you feel miserable. It's quite natural that when your symptoms are at their very worst, you will do things to try to get better. You might take a homeopathic remedy. You might sacrifice a goat and dangle its entrails around your neck. You might bully your GP into giving you antibiotics. (I've listed these in order of increasing ridiculousness.)

Then, when you get better – as you surely will from a cold – you will naturally assume that whatever you did when your symptoms were at their worst must be the reason for your recovery. *Post hoc ergo propter hoc*, and all that. Every time you get a cold from now on, you'll be back at your GP, hassling her for antibiotics, and she'll be saying, 'Look, I don't think this is a very good idea,' but you'll insist, because they worked last time, and community antibiotic resistance will increase, and ultimately old ladies die of MRSA because of this kind of irrationality, but that's another story.*

You can look at regression to the mean more mathematically, if you prefer. On Bruce Forsyth's *Play Your Cards Right*, when Brucey puts a 3 on the board, the audience all shout, 'Higher!' because they know the odds are that the next card is going to be higher than a 3. 'Do you want to go higher or lower than a jack? Higher? Higher?' 'Lower!'

An even more extreme version of 'regression to the mean' is what Americans call the *Sports Illustrated* jinx. Whenever a sportsman appears on the cover of *Sports Illustrated*, goes the story, he is soon to fall from grace. But to get on the cover of the magazine you have to be at the absolute top of your game, one of the best sportsmen in the world; and to be the best in that week, you're probably also having an unusual run of luck. Luck, or 'noise', generally passes, it 'regresses to the mean' by itself, as happens with throws of a die. If you fail to understand that, you start looking for another cause for that regression, and you find … the *Sports Illustrated* jinx.

---

* General practitioners sometimes prescribe antibiotics to demanding patients in exasperation, even though they are ineffective in treating a viral cold, but much research suggests that this is counterproductive, even as a time-saver. In one study, prescribing antibiotics rather than giving advice on self-management for sore throat resulted in an increased overall workload through repeat attendance. It was calculated that if a GP prescribed antibiotics for sore throat to one hundred fewer patients each year, thirty-three fewer would believe that antibiotics were effective, twenty-five fewer would intend to consult with the problem in the future, and ten fewer would come back within the next year. If you were an alternative therapist, or a drug salesman, you could turn those figures on their head and look at how to drum up more trade, not less.

Homeopaths increase the odds of a perceived success in their treatments even further by talking about 'aggravations', explaining that sometimes the correct remedy can make symptoms get worse before they get better, and claiming that this is part of the treatment process. Similarly, people flogging detox will often say that their remedies might make you feel worse at first, as the toxins are extruded from your body: under the terms of these promises, literally anything that happens to you after a treatment is proof of the therapist's clinical acumen and prescribing skill.

So we could go back to our homeopathy fan, and say: 'You feel you get better, I accept that. But perhaps it is because of "regression to the mean", or simply the "natural history" of the disease.' Again, they cannot say 'No' (or at least not with any meaning – they might say it in a tantrum), because they have no possible way of knowing whether they were going to get better anyway, on the occasions when they apparently got better after seeing a homeopath. 'Regression to the mean' might well be the true explanation for their return to health. They simply cannot tell. They can only restate, again, their original statement: 'All I know is, I feel as if it works. I get better when I take homeopathy.'

That may be as far as they want to go. But when someone goes further, and says, 'Homeopathy works,' or mutters about 'science', then that's a problem. We cannot simply decide such things on the basis of one individual's experiences, for the reasons described above: they might be mistaking the placebo effect for a real effect, or mistaking a chance finding for a real one. Even if we had one genuine, unambiguous and astonishing case of a person getting better from terminal cancer, we'd still be careful about using that one person's experience, because sometimes, entirely by chance, miracles really do happen. Sometimes, but not very often.

Over the course of many years, a team of Australian oncologists followed 2,337 terminal cancer patients in palliative care.

They died, on average, after five months. But around 1 per cent of them were still alive after five years. In January 2006 this study was reported in the *Independent*, bafflingly, as:

'Miracle' Cures Shown to Work

Doctors have found statistical evidence that alternative treatments such as special diets, herbal potions and faith healing can cure apparently terminal illness, but they remain unsure about the reasons.

But the point of the study was specifically *not* that there are miracle cures (it didn't look at any such treatments, that was an invention by the newspaper). Instead, it showed something much more interesting: that amazing things simply happen sometimes: people can survive, despite all the odds, for no apparent reason. As the researchers made clear in their own description, claims for miracle cures should be treated with caution, because 'miracles' occur routinely, in 1 per cent of cases by their definition, and *without* any specific intervention. The lesson of this paper is that we cannot reason from one individual's experience, or even that of a handful, selected out to make a point.

So how do we move on? The answer is that we take lots of individuals, a sample of patients who represent the people we hope to treat, with all of their individual experiences, and count them all up. This is clinical academic medical research, in a nutshell, and there's really nothing more to it than that: no mystery, no 'different paradigm', no smoke and mirrors. It's an entirely transparent process, and this one idea has probably saved more lives, on a more spectacular scale, than any other idea you will come across this year.

It is also not a new idea. The first trial appears in the Old Testament, and interestingly, although nutritionism has only recently become what we might call the 'bollocks *du jour*', it was about food. Daniel was arguing with King Nebuchadnezzar's chief eunuch over the Judaean captives' rations. Their diet was rich food and wine, but Daniel wanted his own soldiers to be given only vegetables. The eunuch was worried that they would become worse soldiers if they didn't eat their rich meals, and that whatever could be done to a eunuch to make his life worse might be done to him. Daniel, on the other hand, was willing to compromise, so he suggested the first ever clinical trial:

> And Daniel said unto the guard … 'Submit us to this test for ten days. Give us only vegetables to eat and water to drink; then compare our looks with those of the young men who have lived on the food assigned by the King and be guided in your treatment of us by what you see.'
>
> The guard listened to what they said and tested them for ten days. At the end of ten days they looked healthier and were better nourished than all the young men who had lived on the food assigned them by the King. So the guard took away the assignment of food and the wine they were to drink and gave them only the vegetables.
>
> *Daniel 1: 1–16.*

To an extent, that's all there is to it: there's nothing particularly mysterious about a trial, and if we wanted to see whether homeopathy pills work, we could do a very similar trial. Let's flesh it out. We would take, say, two hundred people going to a homeopathy clinic, divide them randomly into two groups, and let them go through the whole process of seeing the homeopath, being diagnosed, and getting their prescription for whatever the homeopath wants to give them. But at the last minute, without their knowledge, we would switch half of the patients' homeopathic sugar pills, giving them dud sugar pills, that have not been magically potentised by homeopathy. Then, at an appropriate time later, we could measure how many in each group got better.

Speaking with homeopaths, I have encountered a great deal of angst about the idea of measuring, as if this was somehow not a transparent process, as if it forces a square peg into a round hole, because 'measuring' sounds scientific and mathematical. We should pause for just a moment and think about this clearly. Measuring involves no mystery, and no special devices. We ask people if they feel better, and count up the answers.

In a trial – or sometimes routinely in outpatients' clinic – we might ask people to measure their knee pain on a scale of one to ten every day, in a diary. Or to count up the number of pain-free days in a week. Or to measure the effect their fatigue has had on their life that week: how many days they've been able to get out of the house, how far they've been able to walk, how much housework they've been able to do. You can ask about any number of very simple, transparent, and often quite subjective things, because the business of medicine is improving lives, and ameliorating distress.

We might dress the process up a bit, to standardise it, and allow our results to be compared more easily with other research (which is a good thing, as it helps us to get a broader understanding of a condition and its treatment). We might use the 'General Health Questionnaire', for example, because it's a standardised 'tool'; but for all the bluster, the 'GHQ-12', as it is known, is just a simple list of questions about your life and your symptoms.

If anti-authoritarian rhetoric is your thing, then bear this in mind: perpetrating a placebo-controlled trial of an accepted treatment – whether it's an alternative therapy or any form of medicine – is an inherently subversive act. You undermine false certainty, and you deprive doctors, patients and therapists of treatments which previously pleased them.

There is a long history of upset being caused by trials, in medicine as much as anywhere, and all kinds of people will mount all kinds of defences against them. Archie Cochrane, one of the grandfathers of evidence-based medicine, once amusingly described how different groups of surgeons were each earnestly contending that their treatment for cancer was the most effective: it was transparently obvious to them all that their own treatment was the best. Cochrane went so far as to bring a collection of them together in a room, so that they could witness each other's dogged but conflicting certainty, in his efforts to persuade them of the need for trials. Judges, similarly, can be highly resistant to the notion of trialling different forms of sentence for heroin users, believing that they know best in each individual case. These are recent battles, and they are in no sense unique to the world of homeopathy.

So, we take our group of people coming out of a homeopathy clinic, we switch half their pills for placebo pills, and we measure who gets better. That's a placebo-controlled trial of homeopathy pills, and this is not a hypothetical discussion: these trials have been done on homeopathy, and it seems that overall, homeopathy does no better than placebo.

And yet you will have heard homeopaths say that there are positive trials in homeopathy; you may even have seen specific ones quoted. What's going on here? The answer is fascinating, and takes us right to the heart of evidence-based medicine. There are *some* trials which find homeopathy to perform better than placebo, but only some, and they are, in general, trials with 'methodological flaws'. This sounds technical, but all it means is that there are problems in the way the trials were performed, and those problems are so great that they mean the trials are less 'fair tests' of a treatment.

The alternative therapy literature is certainly riddled with incompetence, but flaws in trials are actually very common throughout medicine. In fact, it would be fair to say that all research has some 'flaws', simply because every trial will involve a compromise between what would be ideal, and what is practical or cheap. (What sets the CAM literature apart is, in some respects, the interpretation: medics sometimes know if they're quoting duff papers, and describe the flaws, whereas homeopaths tend to be uncritical of anything positive.)

That is why it's important that research is always published, in full, with its methods and results available for scrutiny. This is a recurring theme in this book, and it's important, because when people make claims based upon their research, we need to be able to decide for ourselves how big the 'methodological flaws' were, and come to our own judgement about whether the results are reliable, whether theirs was a 'fair test'. The things that stop a trial from being fair are, once you know about them, blindingly obvious.

Blinding

One important feature of a good trial is that neither the experimenters nor the patients know if they got the homeopathy sugar pill or the simple placebo sugar pill, because we want to be sure that any difference we measure is the result of the difference between the pills, and not of people's expectations or biases. If the researchers knew which of their beloved patients were having the real and which the placebo pills, they might give the game away – or it might change their assessment of the patient – consciously or unconsciously.

Let's say I'm doing a study on a medical pill designed to reduce high blood pressure. I know which of my patients are having the expensive new blood pressure pill, and which are having the placebo. One of the people on the swanky new blood pressure pills comes in and has a blood pressure reading that is way off the scale, much higher than I would have expected, especially since they're on this expensive new drug. So I recheck their blood pressure, 'just to make sure I didn't make a mistake'. The next result is more normal, so I write that one down, and ignore the high one.

Blood pressure readings are an inexact technique, like ECG interpretation, X-ray interpretation, pain scores, and many other measurements that are routinely used in clinical trials. I go for lunch, entirely unaware that I am calmly and quietly polluting the data, destroying the study, producing inaccurate evidence, and therefore, ultimately, killing people (because our greatest mistake would be to forget that data is used for serious decisions in the very real world, and bad information causes suffering and death).

There are several good examples from recent medical history where a failure to ensure adequate 'blinding', as it is called, has resulted in the entire medical profession being mistaken about which was the better treatment. We had no way of knowing whether keyhole surgery was better than open surgery, for example, until a group of surgeons from Sheffield came along and did a very theatrical trial, in which bandages and decorative fake blood squirts were used, to make sure that nobody could tell which type of operation anyone had received.

Some of the biggest figures in evidence-based medicine got together and did a review of blinding in all kinds of trials of medical drugs, and found that trials with inadequate blinding exaggerated the benefits of the treatments being studied by 17 per cent. Blinding is not some obscure piece of nitpicking, idiosyncratic to pedants like me, used to attack alternative therapies.

Closer to home for homeopathy, a review of trials of acupuncture for back pain showed that the studies which were properly blinded showed a tiny benefit for acupuncture, which was not 'statistically significant' (we'll come back to what that means later). Meanwhile, the trials which were not blinded – the ones where the patients knew whether they were in the treatment group or not – showed a massive, statistically significant benefit for acupuncture. (The placebo control for acupuncture, in case you're wondering, is sham acupuncture, with fake needles, or needles in the 'wrong' places, although an amusing complication is that sometimes one school of acupuncturists will claim that another school's sham needle locations are actually their genuine ones.)

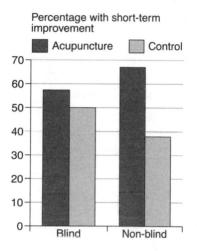

So, as we can see, blinding is important, and not every trial is necessarily any good. You can't just say, 'Here's a trial that shows this treatment works,' because there are good trials, or 'fair tests', and there are bad trials. When doctors and scientists say that a study was methodologically flawed and unreliable, it's not because they're being mean, or trying to maintain the 'hegemony', or to keep the backhanders coming from the pharmaceutical industry: it's because the study was poorly performed – it costs nothing to blind properly – and simply wasn't a fair test.

Randomisation

Let's take this out of the theoretical, and look at some of the trials which homeopaths quote to support their practice. I've got a bog-standard review of trials for homeopathic arnica by Professor Edward Ernst in front of me, which we can go through for examples. We should be absolutely clear that the inadequacies here are not unique, I do not imply malice, and I am not being mean. What we are doing is simply what medics and academics do when they appraise evidence.

So, Hildebrandt *et al.* (as they say in academia) looked at forty-two women taking homeopathic arnica for delayed-onset muscle soreness, and found it performed better than placebo. At first glance this seems to be a pretty plausible study, but if you look closer, you can see there was no 'randomisation' described. Randomisation is another basic concept in clinical trials. We randomly assign patients to the placebo sugar pill group or the homeopathy sugar pill group, because otherwise there is a risk that the doctor or homeopath – consciously or unconsciously – will put patients who they think might do well into the homeopathy group, and the no-hopers into the placebo group, thus rigging the results.

Randomisation is not a new idea. It was first proposed in the seventeenth century by John Baptista van Helmont, a Belgian radical who challenged the academics of his day to test their treatments like blood-letting and purging (based on 'theory') against his own, which he said were based more on clinical experience: 'Let us take out of the hospitals, out of the Camps, or from elsewhere, two hundred, or five hundred poor People, that have Fevers, Pleurisies, etc. Let us divide them into half, let us cast lots, that one half of them may fall to my share, and the other to yours ... We shall see how many funerals both of us shall have.'

It's rare to find an experimenter so careless that they've not randomised the patients at all, even in the world of CAM. But it's surprisingly common to find trials where the method of randomisation is inadequate: they look plausible at first glance, but on closer examination we can see that the experimenters have simply gone through a kind of theatre, as if they were randomising the patients, but still leaving room for them to influence, consciously or unconsciously, which group each patient goes into.

In some inept trials, in all areas of medicine, patients are 'randomised' into the treatment or placebo group by the order in which they are recruited onto the study – the first patient in gets the real treatment, the second gets the placebo, the third the real treatment, the fourth the placebo, and so on. This sounds fair enough, but in fact it's a glaring hole that opens your trial up to possible systematic bias.

Let's imagine there is a patient who the homeopath believes to be a no-hoper, a heart-sink patient who'll never really get better, no matter what treatment he or she gets, and the next place available on the study is for someone going into the 'homeopathy' arm of the trial. It's not inconceivable that the homeopath might just decide – again, consciously or unconsciously – that this particular patient 'probably wouldn't really be interested' in the trial. But if, on the other hand, this no-hoper patient had come into clinic at a time when the next place on the trial was for the placebo group, the recruiting clinician might feel a lot more optimistic about signing them up.

The same goes for all the other inadequate methods of randomisation: by last digit of date of birth, by date seen in clinic, and so on. There are even studies which claim to randomise patients by tossing a coin, but forgive me (and the entire evidence-based medicine community) for worrying that tossing a coin leaves itself just a little bit too open to manipulation. Best of three, and all that. Sorry, I meant best of five. Oh, I didn't really see that one, it fell on the floor.

There are plenty of genuinely fair methods of randomisation, and although they require a bit of nous, they come at no extra financial cost. The classic is to make people call a special telephone number, to where someone is sitting with a computerised randomisation programme (and the experimenter doesn't even do that until the patient is fully signed up and committed to the study). This is probably the most popular method amongst meticulous researchers, who are keen to ensure they are doing a 'fair test', simply because you'd have to be an out-and-out charlatan to mess it up, and you'd have to work pretty hard at the charlatanry too. We'll get back to laughing at quacks in a minute, but right now you are learning about one of the most important ideas of modern intellectual history.

Does randomisation matter? As with blinding, people have studied the effect of randomisation in huge reviews of large numbers of trials, and found that the ones with dodgy methods of randomisation overestimate treatment effects by 41 per cent. In reality, the biggest problem with poor-quality trials is not that they've used an inadequate method of randomisation, it's that they don't tell you *how* they randomised the patients at all. This is a classic warning sign, and often means the trial has been performed badly. Again, I do not speak from prejudice: trials with unclear methods of randomisation overstate treatment effects by 30 per cent, almost as much as the trials with openly rubbish methods of randomisation.

In fact, as a general rule it's always worth worrying when people don't give you sufficient details about their methods and results. As it happens (I promise I'll stop this soon), there have been two landmark studies on whether inadequate information in academic articles is associated with dodgy, overly flattering results, and yes, studies which don't report their methods fully do overstate the benefits of the treatments, by around 25 per cent. Transparency and detail are everything in science. Hildebrandt *et al.*, through no fault of their own, happened to be the peg for this discussion on randomisation (and I am grateful to them for it): they might well have randomised their patients. They might well have done so adequately. But they did not report on it.

Let's go back to the eight studies in Ernst's review article on homeopathic arnica – which we chose pretty arbitrarily – because they demonstrate a phenomenon which we see over and over again with CAM studies: most of the trials were hopelessly

methodologically flawed, and showed positive results for homeopathy; whereas the couple of decent studies – the most 'fair tests'
– showed homeopathy to perform no better than placebo.*

So now you can see, I would hope, that when doctors say a
piece of research is 'unreliable', that's not necessarily a stitch-up;
when academics deliberately exclude a poorly performed study
that flatters homeopathy, or any other kind of paper, from a
systematic review of the literature, it's not through a personal or

---

* So, Pinsent performed a double-blind, placebo-controlled study of fifty-
nine people having oral surgery: the group receiving homeopathic arnica
experienced significantly less pain than the group getting placebo. What you
don't tend to read in the arnica publicity material is that forty one subjects
dropped out of this study. That makes it a fairly rubbish study. It's been shown
that patients who drop out of studies are less likely to have taken their tablets
properly, more likely to have had side-effects, less likely to have got better, and
so on. I am not sceptical about this study because it offends my prejudices, but
because of the high drop-out rate. The missing patients might have been lost
to follow-up because they are dead, for example. Ignoring drop-outs tends to
exaggerate the benefits of the treatment being tested, and a high drop-out rate
is always a warning sign.

The study by Gibson *et al.* did not mention randomisation, nor did it
deign to mention the dose of the homeopathic remedy, or the frequency with
which it was given. It's not easy to take studies very seriously when they are
this thin.

There was a study by Campbell which had thirteen subjects in it (which
means a tiny handful of patients in both the homeopathy and the placebo
groups): it found that homeopathy performed better than placebo (in this
teeny-tiny sample of subjects), but didn't check whether the results were
statistically significant, or merely chance findings.

Lastly, Savage *et al.* did a study with a mere ten patients, finding that
homeopathy was better than placebo; but they too did no statistical analysis of
their results.

These are the kinds of papers that homeopaths claim as evidence to
support their case, evidence which they claim is deceitfully ignored by the
medical profession. All of these studies favoured homeopathy. All deserve to
be ignored, for the simple reason that each was not a 'fair test' of homeopathy,
simply on account of these methodological flaws.

I could go on, through a hundred homeopathy trials, but it's painful
enough already.

moral bias: it's for the simple reason that if a study is no good, if it is not a 'fair test' of the treatments, then it might give unreliable results, and so it should be regarded with great caution.

There is a moral and financial issue here too: randomising your patients properly doesn't cost money. Blinding your patients to whether they had the active treatment or the placebo doesn't cost money. Overall, doing research robustly and fairly does not necessarily require more money, it simply requires that you think before you start. The only people to blame for the flaws in these studies are the people who performed them. In some cases they will be people who turn their backs on the scientific method as a 'flawed paradigm'; and yet it seems their great new paradigm is simply 'unfair tests'.

These patterns are reflected throughout the alternative therapy literature. In general, the studies which are flawed tend to be the ones that favour homeopathy, or any other alternative therapy; and the well-performed studies, where every controllable source of bias and error is excluded, tend to show that the treatments are no better than placebo.

This phenomenon has been carefully studied, and there is an almost linear relationship between the methodological quality of a homeopathy trial and the result it gives. The worse the study – which is to say, the less it is a 'fair test' – the more likely it is to find that homeopathy is better than placebo. Academics conventionally measure the quality of a study using standardised tools like the 'Jadad score', a seven-point tick list that includes things we've been talking about, like 'Did they describe the method of randomisation?' and 'Was plenty of numerical information provided?'

This graph, from Ernst's paper, shows what happens when you plot Jadad score against result in homeopathy trials. Towards the top left, you can see rubbish trials with huge design flaws which triumphantly find that homeopathy is much, much better than placebo. Towards the bottom right, you can see that as the Jadad score tends towards the top mark of 5, as the trials become more of a 'fair test', the line tends towards showing that homeopathy performs no better than placebo.

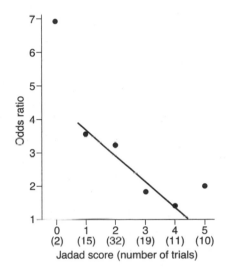

Jadad score (number of trials)

There is, however, a mystery in this graph: an oddity, and the makings of a whodunnit. That little dot on the right-hand edge of the graph, representing the ten best-quality trials, with the highest Jadad scores, stands clearly outside the trend of all the others. This is an anomalous finding: suddenly, only at that end of the graph, there are some good-quality trials bucking the trend and showing that homeopathy is better than placebo.

What's going on there? I can tell you what I think: some of the papers making up that spot are a stitch-up. I don't know which ones, how it happened, or who did it, in which of the ten papers, but that's what I think. Academics often have to couch strong criticism in diplomatic language. Here is Professor Ernst, the man who made that graph, discussing the eyebrow-raising outlier. You might decode his *Yes, Minister* diplomacy, and conclude that he thinks there's been a stitch-up too.

There may be several hypotheses to explain this phenomenon. Scientists who insist that homeopathic remedies are in every way identical to placebos might favour the following. The correlation provided by the four data points (Jadad score 1–4) roughly reflects the truth. Extrapolation of this correlation would lead them to expect that those trials with the least room for bias (Jadad score = 5) show homeopathic remedies are pure placebos. The fact, however, that the average result of the 10 trials scoring 5 points on the Jadad score contradicts this notion, is consistent with the hypothesis that some (by no means all) methodologically astute and highly convinced homeopaths have published results that look convincing but are, in fact, not credible.

But this is a curiosity and an aside. In the bigger picture it doesn't matter, because overall, even including these suspicious studies, the 'meta-analyses' still show, overall, that homeopathy is no better than placebo.

Meta-analyses?

Meta-analysis

This will be our last big idea for a while, and this is one that has saved the lives of more people than you will ever meet. A meta-analysis is a very simple thing to do, in some respects: you just collect all the results from all the trials on a given subject, bung them into one big spreadsheet, and do the maths on that, instead of relying on your own gestalt intuition about all the results from each of your little trials. It's particularly useful when there have been lots of trials, each too small to give a conclusive answer, but all looking at the same topic.

So if there are, say, ten randomised, placebo-controlled trials looking at whether asthma symptoms get better with homeopathy, each of which has a paltry forty patients, you could put them all into one meta-analysis and effectively (in some respects) have a four-hundred-person trial to work with.

In some very famous cases – at least, famous in the world of academic medicine – meta-analyses have shown that a treatment previously believed to be ineffective is in fact rather good, but because the trials that had been done were each too small, individually, to detect the real benefit, nobody had been able to spot it.

As I said, information alone can be life-saving, and one of the greatest institutional innovations of the past thirty years is undoubtedly the Cochrane Collaboration, an international not-for-profit organisation of academics, which produces systematic summaries of the research literature on healthcare research, including meta-analyses.

The logo of the Cochrane Collaboration features a simplified 'blobbogram', a graph of the results from a landmark meta-analysis which looked at an intervention given to pregnant mothers. When people give birth prematurely, as you might expect, the babies are more likely to suffer and die. Some doctors in New Zealand had the idea that giving a short, cheap course of a steroid might help improve outcomes, and seven trials testing this idea were done between 1972 and 1981. Two of them showed some benefit from the steroids, but the remaining five failed to detect any benefit, and because of this, the idea didn't catch on.

THE COCHRANE COLLABORATION®

Eight years later, in 1989, a meta-analysis was done by pooling all this trial data. If you look at the blobbogram in the logo on the previous page, you can see what happened. Each horizontal line represents a single study: if the line is over to the left, it means the steroids were better than placebo, and if it is over to the right, it means the steroids were worse. If the horizontal line for a trial touches the big vertical 'nil effect' line going down the middle, then the trial showed no clear difference either way. One last thing: the longer a horizontal line is, the less certain the outcome of the study was.

Looking at the blobbogram, we can see that there are lots of not-very-certain studies, long horizontal lines, mostly touching the central vertical line of 'no effect'; but they're all a bit over to the left, so they all seem to suggest that steroids *might* be beneficial, even if each study itself is not statistically significant.

The diamond at the bottom shows the pooled answer: that there is, in fact, very strong evidence indeed for steroids reducing the risk – by 30 to 50 per cent – of babies dying from the complications of immaturity. We should always remember the human cost of these abstract numbers: babies died unnecessarily because they were deprived of this life-saving treatment for a decade. They died, *even when there was enough information available to know what would save them*, because that information had not been synthesised together, and analysed systematically, in a meta-analysis.

Back to homeopathy (you can see why I find it trivial now). A landmark meta-analysis was published recently in the *Lancet*. It was accompanied by an editorial titled: 'The End of Homeopathy?' Shang *et al.* did a very thorough meta-analysis of a vast number of homeopathy trials, and they found, overall, adding them all up, that homeopathy performs no better than placebo.

The homeopaths were up in arms. If you mention this meta-analysis, they will try to tell you that it was a stitch-up. What Shang *et al.* did, essentially, like all the previous negative meta-analyses of homeopathy, was to exclude the poorer-quality trials from their analysis.

Homeopaths like to pick out the trials that give them the answer that they want to hear, and ignore the rest, a practice called 'cherry-picking'. But you can also cherry-pick your favourite meta-analyses, or misrepresent them. Shang *et al.* was only the latest in a long string of meta-analyses to show that homeopathy performs no better than placebo. What is truly amazing to me is that despite the negative results of these meta-analyses, homeopaths have continued – right to the top of the profession – to claim that these same meta-analyses *support* the use of homeopathy. They do this by quoting only the result for *all* trials included in each meta-analysis. This figure includes all of the poorer-quality trials. The most reliable figure, you now know, is for the restricted pool of the most 'fair tests', and when you look at those, homeopathy performs no better than placebo. If this fascinates you (and I would be very surprised), then I am currently producing a summary with some colleagues, and you will soon be able to find it online at badscience.net, in all its glorious detail, explaining the results of the various meta-analyses performed on homeopathy.

Clinicians, pundits and researchers all like to say things like 'There is a need for more research,' because it sounds forward-thinking and open-minded. In fact that's not always the case, and it's a little-known fact that this very phrase has been effectively banned from the *British Medical Journal* for many years, on the grounds that it adds nothing: you may say what research is missing, on whom, how, measuring what, and why you want to do it, but the hand-waving, superficially open-minded call for 'more research' is meaningless and unhelpful.

There have been over a hundred randomised placebo-controlled trials of homeopathy, and the time has come to stop. Homeopathy pills work no better than placebo pills, we know that much. But there is room for more interesting research.

People do experience that homeopathy is positive for them, but the action is likely to be in the whole process of going to see a homeopath, of being listened to, having some kind of explanation for your symptoms, and all the other collateral benefits of old-fashioned, paternalistic, reassuring medicine. (Oh, and regression to the mean.)

So we should measure that; and here is the final superb lesson in evidence-based medicine that homeopathy can teach us: sometimes you need to be imaginative about what kinds of research you do, compromise, and be driven by the questions that need answering, rather than the tools available to you.

It is very common for researchers to research the things which interest them, in all areas of medicine; but they can be interested in quite different things from patients. One study actually thought to ask people with osteoarthritis of the knee what kind of research they wanted to be carried out, and the responses were fascinating: they wanted rigorous real-world evaluations of the benefits from physiotherapy and surgery, from educational and coping strategy interventions, and other pragmatic things. They didn't want yet another trial comparing one pill with another, or with placebo.

In the case of homeopathy, similarly, homeopaths want to believe that the power is in the pill, rather than in the whole process of going to visit a homeopath, having a chat and so on. It is crucially important to their professional identity. But I believe that going to see a homeopath is probably a helpful intervention, in some cases, for some people, even if the pills are just placebos. I think patients would agree, and I think it would be an interesting thing to measure. It would be easy, and you would do something called a pragmatic 'waiting-list-controlled trial'.

You take two hundred patients, say, all suitable for homeopathic treatment, currently in a GP clinic, and all willing to be referred on for homeopathy, then you split them randomly into two groups of one hundred. One group gets treated by a homeopath as normal, pills, consultation, smoke and voodoo, on top of whatever other treatment they are having, just like in the real world. The other group just sits on the waiting list. They get treatment as usual, whether that is 'neglect', 'GP treatment' or whatever, but no homeopathy. Then you measure outcomes, and compare who gets better the most.

You could argue that it would be a trivial positive finding, and that it's obvious the homeopathy group would do better; but it's the only piece of research really waiting to be done. This is a 'pragmatic trial'. The groups aren't blinded, but they couldn't possibly be in this kind of trial, and sometimes we have to accept compromises in experimental methodology. It would be a legitimate use of public money (or perhaps money from Boiron, the homeopathic pill company valued at $500 million), but there's nothing to stop homeopaths from just cracking on and doing it for themselves: because despite the homeopaths' fantasies, born out of a lack of knowledge, that research is diffi-cult, magical and expensive, in fact such a trial would be very cheap to conduct.

In fact, it's not really money that's missing from the alterna-tive therapy research community, especially in Britain: it's knowledge of evidence-based medicine, and expertise in how to do a trial. Their literature and debates drip with ignorance, and vitriolic anger at anyone who dares to appraise the trials. Their university courses, as far as they ever even dare to admit what they teach on them (it's all suspiciously hidden away), seem to skirt around such explosive and threatening questions. I've suggested in various places, including at academic conferences, that the single thing that would most improve the quality of evidence in CAM would be funding for a simple, evidence-based medicine hotline, which anyone thinking about running a trial in their clinic could phone up and get advice on how to do it properly, to avoid wasting effort on an 'unfair test' that will rightly be regarded with contempt by all outsiders.

In my pipe dream (I'm completely serious, if you've got the money) you'd need a handout, maybe a short course that people did to cover the basics, so they weren't asking stupid questions, and phone support. In the meantime, if you're a sensible homeopath and you want to do a GP-controlled trial, you could maybe try the badscience website forums, where there are people who might be able to give some pointers (among the childish fighters and trolls …).

But would the homeopaths buy it? I think it would offend their sense of professionalism. You often see homeopaths trying to nuance their way through this tricky area, and they can't quite make their minds up. Here, for example, is a Radio 4 interview, archived in full online, where Dr Elizabeth Thompson (consultant homeopathic physician, and honorary senior lecturer at the Department of Palliative Medicine at the University of Bristol) has a go.

She starts off with some sensible stuff: homeopathy does work, but through non-specific effects, the cultural meaning of the process, the therapeutic relationship, it's not about the pills, and so on. She practically comes out and says that homeopathy is all about cultural meaning and the placebo effect. 'People have wanted to say homeopathy is like a pharmaceutical compound,' she says, 'and it isn't, it is a complex intervention.'

Then the interviewer asks: 'What would you say to people who go along to their high street pharmacy, where you can buy homeopathic remedies, they have hay fever and they pick out a hay-fever remedy, I mean presumably that's not the way it works?' There is a moment of tension. Forgive me, Dr Thompson, but I felt you didn't want to say that the pills work, as pills, in isolation, when you buy them in a shop: apart from anything else, you'd already said that they don't.

But she doesn't want to break ranks and say the pills don't work, either. I'm holding my breath. How will she do it? Is there a linguistic structure complex enough, passive enough, to negotiate through this? If there is, Dr Thompson doesn't find it: 'They might flick through and they might just be spot-on … [but] you've got to be very lucky to walk in and just get the right remedy.' So the power is, and is not, in the pill: 'P, and not-P', as philosophers of logic would say.

If they can't finesse it with the 'power is not in the pill' paradox, how else do the homeopaths get around all this negative data? Dr Thompson – from what I have seen – is a fairly clear-thinking and civilised homeopath. She is, in many respects, alone. Homeopaths have been careful to keep themselves outside of the civilising environment of the university, where the influence and questioning of colleagues can help to refine ideas, and weed out the bad ones. In their rare forays, they enter them secretively, walling themselves and their ideas off from criticism or review, refusing to share even what is in their exam papers with outsiders.

It is rare to find a homeopath engaging on the issue of the evidence, but what happens when they do? I can tell you. They get angry, they threaten to sue, they scream and shout at you at meetings, they complain spuriously and with ludicrous misrepresentations – time-consuming to expose, of course, but that's the point of harassment – to the Press Complaints Commission and your editor, they send hate mail, and accuse you repeatedly of somehow being in the pocket of big pharma (falsely, although you start to wonder why you bother having principles when faced with this kind of behaviour). They bully, they smear, to the absolute top of the profession, and they do anything they can in a desperate bid to *shut you up*, and avoid having a discussion about the evidence. They have even been known to threaten violence (I won't go into it here, but I manage these issues extremely seriously).

I'm not saying I don't enjoy a bit of banter. I'm just pointing out that you don't get anything quite like this in most other fields, and homeopaths, among all the people in this book, with the exception of the odd nutritionist, seem to me to be a uniquely angry breed. Experiment for yourself by chatting with them about evidence, and let me know what you find.

By now your head is hurting, because of all those mischievous, confusing homeopaths and their weird, labyrinthine defences: you need a lovely science massage. Why is evidence so complicated? Why do we need all of these clever tricks, these special research paradigms? The answer is simple: the world is much more complicated than simple stories about pills making people get better. We are human, we are irrational, we have foibles, and the power of the mind over the body is greater than anything you have previously imagined.

5

The Placebo Effect

For all the dangers of CAM, to me the greatest disappointment is the way it distorts our understanding of our bodies. Just as the Big Bang theory is far more interesting than the creation story in Genesis, so the story that science can tell us about the natural world is far more interesting than any fable about magic pills concocted by an alternative therapist. To redress that balance, I'm offering you a whirlwind tour of one of the most bizarre and enlightening areas of medical research: the relationship between our bodies and our minds, the role of meaning in healing, and in particular the 'placebo effect'.

Much like quackery, placebos became unfashionable in medicine once the biomedical model started to produce tangible results. An editorial in 1890 sounded its death knell, describing the case of a doctor who had injected his patient with water instead of morphine: she recovered perfectly well, but then discovered the deception, disputed the bill in court, and won. The editorial was a lament, because doctors have known that reassurance and a good bedside manner can be very effective for as long as medicine has existed. 'Shall [the placebo] never again have an opportunity of exerting its wonderful psychological effects as faithfully as one of its more toxic conveners?' asked the *Medical Press* at the time.

Luckily, its use survived. Throughout history, the placebo effect has been particularly well documented in the field of pain, and some of the stories are striking. Henry Beecher, an American anaesthetist, wrote about operating on a soldier with horrific injuries in a World War II field hospital, using salt water because the morphine was all gone, and to his astonishment the patient was fine. Peter Parker, an American missionary, described performing surgery without anaesthesia on a Chinese patient in the mid-nineteenth century: after the operation, she 'jumped upon the floor', bowed, and walked out of the room as if nothing had happened.

Theodor Kocher performed 1,600 thyroidectomies without anaesthesia in Berne in the 1890s, and I take my hat off to a man who can do complicated neck operations on conscious patients. Mitchel in the early twentieth century was performing full amputations and mastectomies, entirely without anaesthesia; and surgeons from before the invention of anaesthesia often described how some patients could tolerate knife cutting through muscle, and saw cutting through bone, perfectly awake, and without even clenching their teeth. You might be tougher than you think.

This is an interesting context in which to remember two televised stunts from 2006. The first was a rather melodramatic operation 'under hypnosis' on Channel 4: 'We just want to start the debate on this important medical issue,' explained the production company Zigzag, known for making shows like *Mile High Club* and *Streak Party*. The operation, a trivial hernia repair, was performed with medical drugs but at a reduced dose, and treated as if it was a medical miracle.

The second was in *Alternative Medicine: The Evidence*, a rather gushing show on BBC2 presented by Kathy Sykes ('Professor of the Public Understanding of Science'). This series was the subject of a successful complaint at the highest level, on account of it misleading the audience. Viewers believed they had seen a patient having chest surgery with only acupuncture as anaesthesia: in fact this was not the case, and once again the patient had received an array of conventional medications to allow the operation to be performed.*

When you consider these misleading episodes alongside the reality – that operations have frequently been performed with *no* anaesthetics, *no* placebos, *no* alternative therapists, *no* hypnotists and *no* TV producers – these televised episodes suddenly feel rather less dramatic.

But these are just stories, and the plural of anecdote is not data. Everyone knows about the power of the mind – whether it's stories of mothers enduring biblical pain to avoid dropping a boiling kettle on their baby, or people lifting cars off their girl-friend like the Incredible Hulk – but devising an experiment that teases the psychological and cultural benefits of a treat-ment away from the biomedical effects is trickier than you might think. After all, what do you compare a placebo against? Another placebo? Or no treatment at all?

The placebo on trial

In most studies we don't have a 'no treatment' group to compare both the placebo and the drug against, and for a very good ethical reason: if your patients are ill, you shouldn't be

---

* The series also featured a brain-imaging experiment on acupuncture, funded by the BBC, and one of the scientists involved came out afterwards to complain not only that the results had been overinterpreted (which you would expect from the media, as we will see), but moreover, that the pressure from the funder – that is to say, the BBC – to produce a positive result was overwhelming. This is a perfect example of the things which you do *not* do in science, and the fact that it was masterminded by a 'Professor of the Public Understanding of Science' goes some way towards explaining why we are in such a dismal position today. The programme was defended by the BBC in a letter with ten academic signatories. Several of these signatories have since said they did not sign the letter. The mind really does boggle.

leaving them untreated simply because of your own mawkish interest in the placebo effect. In fact, in most cases today it is considered wrong even to use a placebo in a trial: whenever possible you should compare your new treatment against the best pre-existing, current treatment.

This is not just for ethical reasons (although it is enshrined in the Declaration of Helsinki, the international ethics bible). Placebo-controlled trials are also frowned upon by the evidence-based medicine community, because they know it's an easy way to cook the books and get easy positive trial data to support your company's big new investment. In the real world of clinical practice, patients and doctors aren't so interested in whether a new drug works better than *nothing*, they're interested in whether it works *better than the best treatment they already have.*

There have been occasions in medical history where researchers were more cavalier. The Tuskegee Syphilis Study, for example, is one of America's most shaming hours, if it is possible to say such a thing these days: 399 poor, rural African-American men were recruited by the US Public Health Service in 1932 for an observational study to see what happened if syphilis was left, very simply, untreated. Astonishingly, the study ran right through to 1972. In 1949 penicillin was introduced as an effective treatment for syphilis. These men did not receive that drug, nor did they receive Salvarsan, nor indeed did they receive an apology until 1997, from Bill Clinton.

If we don't want to do unethical scientific experiments with 'no treatment' groups on sick people, how else can we determine the size of the placebo effect on modern illnesses? Firstly, and rather ingeniously, we can compare one placebo with another.

The first experiment in this field was a meta-analysis by Daniel Moerman, an anthropologist who has specialised in the placebo effect. He took the trial data from placebo-controlled trials of gastric ulcer medication, which was his first cunning move, because gastric ulcers are an excellent thing to study: their presence or absence is determined very objectively, with a gastroscopy camera passed down into the stomach, to avoid any doubt.

Moerman took only the placebo data from these trials, and then, in his second ingenious move, from all of these studies, of all the different drugs, with their different dosing regimes, he took the ulcer healing rate from the placebo arm of trials where the 'placebo' treatment was two sugar pills a day, and compared that with the ulcer healing rate in the placebo arm of trials where the placebo was four sugar pills a day. He found, spectacularly, that four sugar pills are better than two (these findings have also been replicated in a different dataset, for those who are switched on enough to worry about the replicability of important clinical findings).

What the treatment looks like

So four pills are better than two: but how can this be? Does a placebo sugar pill simply exert an effect like any other pill? Is there a dose–response curve, as pharmacologists would find for any other drug? The answer is that the placebo effect is about far more than just the pill: it is about the cultural meaning of the treatment. Pills don't simply manifest themselves in your stomach: they are given in particular ways, they take varying forms, and they are swallowed with expectations, all of which have an impact on a person's beliefs about their own health, and in turn, on outcome. Homeopathy is, for example, a perfect example of the value in ceremony.

I understand this might well seem improbable to you, so I've corralled some of the best data on the placebo effect into one place, and the challenge is this: see if you can come up with a better explanation for what is, I guarantee, a seriously strange set of experimental results.

First up, Blackwell [1972] did a set of experiments on fifty-seven college students to determine the effect of colour – as well as the number of tablets – on the effects elicited. The subjects were sitting through a boring hour-long lecture, and were given either one or two pills, which were either pink or blue. They were told that they could expect to receive either a stimulant or a sedative. Since these were psychologists, and this was back when you could do whatever you wanted to your subjects – even lie to them – the treatment that *all* the students received consisted simply of sugar pills, but of different colours.

Afterwards, when they measured alertness – as well as any subjective effects – the researchers found that two pills were more effective than one, as we might have expected (and two pills were better at eliciting side-effects too). They also found that colour had an effect on outcome: the pink sugar tablets were better at maintaining concentration than the blue ones. Since colours in themselves have no intrinsic pharmacological properties, the difference in effect could only be due to the cultural meanings of pink and blue: pink is alerting, blue is cool. Another study suggested that Oxazepam, a drug similar to Valium (which was once unsuccessfully prescribed by our GP for me as a hyperactive child) was more effective at treating anxiety in a green tablet, and more effective for depression when yellow.

Drug companies, more than most, know the benefits of good branding: they spend more on PR, after all, than they do on research and development. As you'd expect from men of action with large houses in the country, they put these theoretical ideas into practice: so Prozac, for example, is white and blue; and in case you think I'm cherry-picking here, a survey of the colour of pills currently on the market found that stimulant medication tends to come in red, orange or yellow tablets, while antidepressants and tranquillisers are generally blue, green or purple.

Issues of form go much deeper than colour. In 1970 a sedative – chlordiazepoxide – was found to be more effective in capsule form than pill form, even for the very same drug, in the very same dose: capsules at the time felt newer, somehow, and more sciencey. Maybe you've caught yourself splashing out and paying extra for ibuprofen capsules in the chemist's.

Route of administration has an effect as well: salt-water injections have been shown in three separate experiments to be more effective than sugar pills for blood pressure, for headaches and for postoperative pain, not because of any physical benefit of salt-water injection over sugar pills – there isn't one – but because, as everyone knows, an injection is a much more dramatic intervention than just taking a pill.

Closer to home for the alternative therapists, the *BMJ* recently published an article comparing two different placebo treatments for arm pain, one of which was a sugar pill, and one of which was a 'ritual', a treatment modelled on acupuncture: the trial found that the more elaborate placebo ritual had a greater benefit.

But the ultimate testament to the social construction of the placebo effect must be the bizarre story of packaging. Pain is an area where you might suspect that expectation would have a particularly significant effect. Most people have found that they can take their minds off pain – to at least some extent – with distraction, or have had a toothache which got worse with stress.

Branthwaite and Cooper did a truly extraordinary study in 1981, looking at 835 women with headaches. It was a four-armed study, where the subjects were given either aspirin or placebo pills, and these pills in turn were packaged either in blank, bland, neutral boxes, or in full, flashy, brand-name packaging. They found – as you'd expect – that aspirin had more of an effect on headaches than sugar pills; but more than that, they found that the packaging itself had a beneficial effect, enhancing the benefit of both the placebo and the aspirin.

People I know still insist on buying brand-name painkillers. As you can imagine, I've spent half my life trying to explain to them why this is a waste of money: but in fact the paradox of Branthwaite and Cooper's experimental data is that they were right all along. Whatever pharmacology theory tells you, that brand-named version *is* better, and there's just no getting away from it. Part of that might be the cost: a recent study looking at pain caused by electric shocks showed that a pain relief treatment was stronger when subjects were told it cost $2.50 than when they were told it cost 10c. (And a paper currently in press shows that people are more likely to take advice when they have paid for it.)

It gets better – or worse, depending on how you feel about your world view slipping sideways. Montgomery and Kirsch [1996] told college students they were taking part in a study on a new local anaesthetic called 'trivaricaine'. Trivaricaine is brown, you paint it on your skin, it smells like a medicine, and it's so potent you have to wear gloves when you handle it: or that's what they implied to the students. In fact it's made of water, iodine and thyme oil (for the smell), and the experimenter (who also wore a white coat) was only using rubber gloves for a sense of theatre. None of these ingredients will affect pain.

The trivaricaine was painted onto one or other of the subjects' index fingers, and the experimenters then applied painful pressure with a vice. One after another, in varying orders, pain was applied, trivaricaine was applied, and as you would expect by now, the subjects reported less pain, and less unpleasantness, for the fingers that were pre-treated with the amazing trivaricaine. This is a placebo effect, but the pills have gone now.

It gets stranger. Sham ultrasound is beneficial for dental pain, placebo operations have been shown to be beneficial in knee pain (the surgeon just makes fake keyhole surgery holes in the side and mucks about for a bit as if he's doing something useful), and placebo operations have even been shown to improve angina.

That's a pretty big deal. Angina is the pain you get when there's not enough oxygen getting to your heart muscle for the work it's doing. That's why it gets worse with exercise: because you're demanding more work from the heart muscle. You might get a similar pain in your thighs after bounding up ten flights of stairs, depending on how fit you are.

Treatments that help angina usually work by dilating the blood vessels to the heart, and a group of chemicals called nitrates are used for this purpose very frequently. They relax the smooth muscle in the body, which dilates the arteries so more blood can get through (they also relax other bits of smooth muscle in the body, including your anal sphincter, which is why a variant is sold as 'liquid gold' in sex shops).

In the 1950s there was an idea that you could get blood vessels in the heart to grow back, and thicker, if you tied off an artery on the front of the chest wall that wasn't very important, but which branched off the main heart arteries. The idea was that this would send messages back to the main branch of the artery, telling it that more artery growth was needed, so the body would be tricked.

Unfortunately this idea turned out to be nonsense, but only after a fashion. In 1959 a placebo-controlled trial of the operation was performed: in some operations they did the whole thing properly, but in the 'placebo' operations they went through the motions but didn't tie off any arteries. It was found that the placebo operation was just as good as the real one – people seemed to get a bit better in both cases, and there was little difference between the groups – but the most strange thing about the whole affair was that nobody made a fuss at the time: the real operation wasn't any better than a sham operation, sure, but how could we explain the fact that people had been sensing an improvement from the operation for a very long time? Nobody thought of the power of placebo. The operation was simply binned.

That's not the only time a placebo benefit has been found at the more dramatic end of the medical spectrum. A Swedish study in the late 1990s showed that patients who had pacemakers installed, but not switched on, did better than they were doing before (although they didn't do as well as people with working pacemakers inside them, to be clear). Even more recently, one study of a very hi-tech 'angioplasty' treatment, involving a large and sciencey-looking laser catheter, showed that sham treatment was almost as effective as the full procedure.

'Electrical machines have great appeal to patients,' wrote Dr Alan Johnson in the *Lancet* in 1994 about this trial, 'and recently anything to do with the word LASER attached to it has caught the imagination.' He's not wrong. I went to visit Lilias Curtin once (she's Cherie Booth's alternative therapist), and she did Gem Therapy on me, with a big shiny science machine that shone different-coloured beams of light onto my chest. It's hard not to see the appeal of things like Gem Therapy in the context of the laser catheter experiment. In fact, the way the evidence is stacking up, it's hard not to see all the claims of alternative therapists, for all their wild, wonderful, authoritative and empathic interventions, in the context of this chapter.

In fact, even the lifestyle gurus get a look in, in the form of an elegant study which examined the effect of simply being told that you are doing something healthy. Eighty-four female room attendants working in various hotels were divided into two groups: one group was told that cleaning hotel rooms is 'good exercise' and 'satisfies the Surgeon General's recommendations for an active lifestyle', along with elaborate explanations of how and why; the 'control' group did not receive this cheering information, and just carried on cleaning hotel rooms. Four weeks later, the 'informed' group perceived themselves to be getting significantly more exercise than before, and showed a significant decrease in weight, body fat, waist-to-hip ratio and body mass index, but amazingly, both groups were still reporting the same amount of activity.*

What the doctor says

If you can believe fervently in your treatment, even though controlled tests show that it is quite useless, then your results are much better, your patients are much better, and your income is much better too. I believe this accounts for the remarkable success of some of the less gifted, but more credulous members of our profession, and also for the violent dislike of statistics and controlled tests which fashionable and successful doctors are accustomed to display.

Richard Asher, Talking Sense, *Pitman Medical, London, 1972*

As you will now be realising, in the study of expectation and belief, we can move away from pills and devices entirely. It turns out, for example, that what the doctor says, and what the doctor believes, both have an effect on healing. If that sounds obvious, I should say they have an effect which has been measured, elegantly, in carefully designed trials.

Gryll and Katahn [1978] gave patients a sugar pill before a dental injection, but the doctors who were handing out the pill gave it in one of two different ways: either with an outrageous oversell ('This is a recently developed pill that's been shown to be very effective … effective almost immediately …'); or down-played, with an undersell ('This is a recently developed pill … personally I've not found it to be very effective …'). The pills which were handed out with the positive message were associated with less fear, less anxiety and less pain.

---

* I agree: this is a bizarre and outrageous experimental finding, and if you have a good explanation for how it might have come about, the world would like to hear from you. Follow the reference, read the full paper online and start a blog, or write a letter to the journal that published it.

Even if he says nothing, what the doctor knows can affect treatment outcomes: the information leaks out, in mannerisms, affect, eyebrows and nervous smiles, as Gracely [1985] demonstrated with a truly ingenious experiment, although understanding it requires a tiny bit of concentration.

He took patients having their wisdom teeth removed, and split them randomly into three treatment groups: they would have either salt water (a placebo that does 'nothing', at least not physiologically), or fentanyl (an excellent opiate painkiller, with a black-market retail value to prove it), or naloxone (an opiate receptor blocker that would actually increase the pain).

In all cases the doctors were blinded to which of the three treatments they were giving to each patient: but Gracely was *really* studying the effect of his doctors' beliefs, so the groups were further divided in half again. In the first group, the doctors giving the treatment were told, truthfully, that they could be administering either placebo, or naloxone, or the pain-relieving fentanyl: this group of doctors knew there was a chance that they were giving something that would reduce pain.

In the second group, the doctors were lied to: they were told they were giving either placebo or naloxone, two things that could only do nothing, or actively make the pain worse. But in fact, without the doctors' knowledge, some of their patients were actually getting the pain-relieving fentanyl. As you would expect by now, just through manipulating what the *doctors believed* about the injection they were giving, even though they were forbidden from vocalising their beliefs to the patients, there was a difference in outcome between the two groups: the first group experienced significantly less pain. This difference had nothing to do with what actual medicine was being given, or even with what information the patients knew: it was entirely down to what the doctors knew. Perhaps they winced when they gave the injection. I think you might have.

'Placebo explanations'

Even if they do nothing, doctors, by their manner alone, can reassure. And even reassurance can in some senses be broken down into informative constituent parts. In 1987, Thomas showed that simply giving a diagnosis – even a fake 'placebo' diagnosis – improved patient outcomes. Two hundred patients with abnormal symptoms, but no signs of any concrete medical diagnosis, were divided randomly into two groups. The patients in one group were told, 'I cannot be certain of what the matter is with you,' and two weeks later only 39 per cent were better; the other group were given a firm diagnosis, with no messing about, and confidently told they would be better within a few days. Sixty-four per cent of that group got better in two weeks.

This raises the spectre of something way beyond the placebo effect, and cuts even further into the work of alternative therapists: because we should remember that alternative therapists don't just give placebo treatments, they also give what we might call 'placebo explanations' or 'placebo diagnoses': ungrounded, unevidenced, often fantastical assertions about the nature of the patient's disease, involving magical properties, or energy, or supposed vitamin deficiencies, or 'imbalances', which the therapist claims uniquely to understand.

And here, it seems that this 'placebo' explanation – even if grounded in sheer fantasy – can be beneficial to a patient, although interestingly, perhaps not without collateral damage, and it must be done delicately: assertively and authoritatively giving someone access to the sick role can also reinforce destructive illness beliefs and behaviours, unnecessarily medicalise symptoms like aching muscles (which for many people are everyday occurrences), and militate against people getting on with life and getting better. It's a very tricky area.

I could go on. In fact there has been a huge amount of research into the value of a good therapeutic relationship, and the general finding is that doctors who adopt a warm, friendly and reassuring manner are more effective than those who keep consultations formal and do not offer reassurance. In the real world, there are structural cultural changes which make it harder and harder for a medical doctor to maximise the therapeutic benefit of a consultation. Firstly, there is the pressure on time: a GP can't do much in a six-minute appointment.

But more than these practical restrictions, there have also been structural changes in the ethical presumptions made by the medical profession, which make reassurance an increasingly outré business. A modern medic would struggle to find a form of words that would permit her to hand out a placebo, for example, and this is because of the difficulty in resolving two very different ethical principles: one is our obligation to heal our patients as effectively as we can; the other is our obligation not to tell them lies. In many cases the prohibition on reassurance and smoothing over worrying facts has been formalised, as the doctor and philosopher Raymond Tallis recently wrote, beyond what might be considered proportionate: 'The drive to keep patients fully informed has led to exponential increases in the formal requirements for consent that only serve to confuse and frighten patients while delaying their access to needed medical attention.'

I don't want to suggest for one moment that historically this was the wrong call. Surveys show that patients want their doctors to tell them the truth about diagnoses and treatments (although you have to take this kind of data with a pinch of salt, because surveys also say that doctors are the most trusted of all public figures, and journalists are the least trusted, but that doesn't seem to be the lesson from the media's MMR hoax).

What is odd, perhaps, is how the primacy of patient autonomy and informed consent over efficacy – which is what we're talking about here – was presumed, but not actively discussed within the medical profession. Although the authoritative and paternalistic reassurance of the Victorian doctor who 'blinds with science' is a thing of the past in medicine, the success of the alternative therapy movement – whose practitioners mislead, mystify and blind their patients with sciencey-sounding 'authoritative' explanations, like the most patronising Victorian doctor imaginable – suggests that there may still be a market for that kind of approach.

About a hundred years ago, these ethical issues were carefully documented by a thoughtful native Canadian Indian called Quesalid. Quesalid was a sceptic: he thought shamanism was bunk, that it only worked through belief, and he went under-cover to investigate this idea. He found a shaman who was willing to take him on, and learned all the tricks of the trade, including the classic performance piece where the healer hides a tuft of down in the corner of his mouth, and then, sucking and heaving, right at the peak of his healing ritual, brings it up, covered in blood from where he has discreetly bitten his lip, and solemnly presents it to the onlookers as a pathological speci-men, extracted from the body of the afflicted patient.

Quesalid had proof of the fakery, he knew the trick as an insider, and was all set to expose those who carried it out; but as part of his training he had to do a bit of clinical work, and he was summoned by a family 'who had dreamed of him as their saviour' to see a patient in distress. He did the trick with the tuft, and was appalled, humbled and amazed to find that his patient got better.

Although he continued to maintain a healthy scepticism about most of his colleagues, Quesalid, to his own surprise perhaps, went on to have a long and productive career as a healer and shaman. The anthropologist Claude Lévi-Strauss, in his paper 'The Sorcerer and his Magic', doesn't quite know what to make of it: 'but it is evident that he carries on his craft consci-entiously, takes pride in his achievements, and warmly defends the technique of the bloody down against all rival schools. He seems to have completely lost sight of the fallaciousness of the technique which he had so disparaged at the beginning.'

Of course, it may not even be necessary to deceive your patient in order to maximise the placebo effect: a classic study from 1965 – albeit small and without a control group – gives a small hint of what might be possible here. They gave a pink placebo sugar pill three times a day to 'neurotic' patients, with good effect, and the explanation given to the patients was startlingly clear about what was going on:

> A script was prepared and carefully enacted as follows: 'Mr. Doe … we have a week between now and your next appointment, and we would like to do something to give you some relief from your symptoms. Many different kinds of tranquilizers and similar pills have been used for conditions such as yours, and many of them have helped. Many people with your kind of condition have also been helped by what are sometimes called 'sugar pills', and we feel that a so-called sugar pill may help you, too. Do you know what a sugar pill is? A sugar pill is a pill with no medicine in it at all. I think this pill will help you as it has helped so many others. Are you willing to try this pill?'
>
> The patient was then given a supply of placebo in the form of pink capsules contained in a small bottle with a label showing the name of the Johns Hopkins Hospital. He was instructed to take the capsules quite regularly, one capsule three times a day at each meal time.

The patients improved considerably. I could go on, but this all sounds a bit wishy-washy: we all know that pain has a strong psychological component. What about the more robust stuff: something more counterintuitive, something more … sciencey?

Dr Stewart Wolf took the placebo effect to the limit. He took two women who were suffering with nausea and vomiting, one of them pregnant, and told them he had a treatment which would improve their symptoms. In fact he passed a tube down into their stomachs (so that they wouldn't taste the revolting bitterness) and administered ipecac, a drug that which should actually *induce* nausea and vomiting.

Not only did the patients' symptoms improve, but their gastric contractions – which ipecac should worsen – were *reduced*. His results suggest – albeit it in a very small sample – that a drug could be made to have the opposite effect to what you would predict from the pharmacology, simply by manipulating people's expectations. In this case, the placebo effect outgunned even the pharmacological influences.

More than molecules?

So is there any research from the basic science of the laboratory bench to explain what's happening when we take a placebo? Well, here and there, yes, although they're not easy experiments to do. It's been shown, for example, that the effects of a real drug in the body can sometimes be induced by the placebo 'version', not only in humans, but also in animals. Most drugs for Parkinson's disease work by increasing dopamine release: patients receiving a placebo treatment for Parkinson's disease, for example, showed extra dopamine release in the brain.

Zubieta [2005] showed that subjects who are subjected to pain, and then given a placebo, release more endorphins than people who got nothing. (I feel duty bound to mention that I'm a bit dubious about this study, because the people on placebo also endured more painful stimuli, which is another reason why they might have had higher endorphins: consider this a small window into the wonderful world of interpreting uncertain data.)

If we delve further into theoretical work from the animal kingdom, we find that animals' immune systems can be conditioned to respond to placebos, in exactly the same way that Pavlov's dog began to salivate in response to the sound of a bell.

Researchers have measured immune system changes in dogs using just flavoured sugar water, once that flavoured water has been associated with immunosuppression, by administering it repeatedly alongside cyclophosphamide, a drug that suppresses the immune system.

A similar effect has been demonstrated in humans, when the researchers gave healthy subjects a distinctively flavoured drink at the same time as cyclosporin A (a drug which measurably reduces your immune function). Once the association was set up with sufficient repetition, they found that the flavoured drink on its own could induce modest immune suppression. Researchers have even managed to elicit an association between sherbet and natural killer cell activity.

What does this all mean for you and me?

People have tended to think, rather pejoratively, that if your pain responds to a placebo, that means it's 'all in the mind'. From survey data, even doctors and nurses buy into this canard. An article from the *Lancet* in 1954 – another planet in terms of how doctors spoke about patients – states that 'for some unintelligent or inadequate patients, life is made easier by a bottle of medicine to comfort the ego'.

This is wrong. It's no good trying to exempt yourself, and pretend that this is about other people, because we all respond to the placebo. Researchers have tried hard in experiments and surveys to characterise 'placebo responders', but the results overall come out like a horoscope that could apply to everybody: 'placebo responders' have been found to be more extroverted but more neurotic, more well-adjusted but more antagonistic, more socially skilled, more belligerent but more acquiescent, and so on. The placebo responder is everyman. You are a placebo responder. Your body plays tricks on your mind. You cannot be trusted.

How do we draw all this together? Moerman reframes the placebo effect as the 'meaning response': 'the psychological and physiological effects of meaning in the treatment of illness', and it's a compelling model. He has also performed one of the most impressive quantitative analyses of the placebo effect, and how it changes with context, again on stomach ulcers. As we've said before, this is an excellent disease to study, because ulcers are prevalent and treatable, but most importantly because treatment success can be unambiguously recorded by having a look down there with a gastroscope.

Moerman examined 117 studies of ulcer drugs from between 1975 and 1994, and found, astonishingly, that they interact in a way you would never have expected: culturally, rather than pharmacodynamically. Cimetidine was one of the first ulcer drugs on the market, and it is still in use today: in 1975, when it was new, it eradicated 80 per cent of ulcers, on average, in the various different trials. As time passed, however, the success rate of cimetidine deteriorated to just 50 per cent. Most interestingly, this deterioration seems to have occurred particularly after the introduction of ranitidine, a competing and supposedly superior drug, onto the market five years later. So the self-same drug became less effective with time, as new drugs were brought in.

There are a lot of possible interpretations of this. It's possible, of course, that it was a function of changing research protocols. But a highly compelling possibility is that the older drugs became less effective after new ones were brought in because of deteriorating medical belief in them. Another study from 2002 looked at seventy-five trials of antidepressants over the previous twenty years, and found that the response to placebo has increased significantly in recent years (as has the response to medication), perhaps as our expectations of those drugs have increased.

Findings like these have important ramifications for our view of the placebo effect, and for all of medicine, since it may be a potent universal force: we must remember, specifically, that the placebo effect – or the 'meaning effect' – is *culturally specific*.

Brand-name painkillers might be better than blank-box painkillers over here, but if you went and found someone with toothache in 6000 BC, or up the Amazon in 1880, or dropped in on Soviet Russia during the 1970s, where nobody had seen the TV advert with the attractive woman wincing from a pulsing red orb of pain in her forehead, who swallows the painkiller, and then the smooth, reassuring blue suffuses her body ... in a world without those cultural preconditions to set up the dominoes, you would expect aspirin to do the same job no matter what box it came out of.

This also has interesting implications for the transferability of alternative therapies. The novelist Jeanette Winterson, for example, has written in *The Times* trying to raise money for a project to treat AIDS sufferers in Botswana – where 48 per cent of the population is HIV positive – with homeopathy. We must put aside the irony here of taking homeopathy to a country that has been engaged in a water war with neighbouring Namibia; and we must also let lie the tragedy of Botswana's devastation by AIDS, which is so phenomenal – I'll say it again: *half the population are HIV positive* – that if it is not addressed rapidly and robustly the entire economically active portion of the population could simply cease to exist, leaving what would be effectively a non-country.

Leaving aside all this tragedy, what's interesting for our purposes is the idea that you could take your Western, individualistic, patient-empowering, anti-medical-establishment and very culturally specific placebo to a country with so little healthcare infrastructure, and expect it to work all the same. The greatest irony of all is that if homeopathy has any benefits at all for AIDS sufferers in Botswana, it may be through its implicit association with the white-coat Western medicine which so many African countries desperately need.

So if you go off now and chat to an alternative therapist about the contents of this chapter – which I very much hope you will –

what will you hear? Will they smile, nod, and agree that their rituals have been carefully and elaborately constructed over many centuries of trial and error to elicit the best placebo response possible? That there are more fascinating mysteries in the true story of the relationship between body and mind than any fanciful notion of quantum energy patterns in a sugar pill?

To me, this is yet another example of a fascinating paradox in the philosophy of alternative therapists: when they claim that their treatments are having a specific and measurable effect on the body, through specific technical mechanisms rather than ritual, they are championing a very old-fashioned and naïve form of biological reductionism, where the mechanics of their interventions, rather than the relationship and the ceremony, have the positive effect on healing. Once again, it's not just that they have no evidence for their claims about how their treatments work: it's that their claims are mechanistic, intellectually disappointing, and simply less interesting than the reality.

An ethical placebo?

But more than anything, the placebo effect throws up fascinating ethical quandaries and conflicts around our feelings on pseudoscience. Let's take our most concrete example so far: are the sugar pills of homeopathy exploitative, if they work only as a placebo? A pragmatic clinician could only consider the value of a treatment by considering it in context.

Here is a clear example of the benefits of placebo. During the nineteenth-century cholera epidemic, deaths were occurring in the London Homeopathic Hospital at just one third of the rate as in the Middlesex Hospital, but a placebo effect is unlikely to be all that beneficial in this condition. The reason for homeopathy's success in this case is more interesting: at the time, nobody could treat cholera. So while hideous medical practices such as blood-letting were actively harmful, the homeopaths' treatments at least did nothing either way.

Today, similarly, there are often situations where people want treatment, but medicine has little to offer – lots of back pain, stress at work, medically unexplained fatigue and most common colds, to give just a few examples. Going through a theatre of medical treatment, and trying every medication in the book, will give you only side-effects. A sugar pill in these circumstances seems a very sensible option, as long as it can be administered cautiously, and ideally with a minimum of deceit.

But just as homeopathy has unexpected benefits, so it can have unexpected side-effects. Believing in things which have no evidence carries its own corrosive intellectual side-effects, just as prescribing a pill in itself carries risks: it medicalises problems, as we will see, it can reinforce destructive beliefs about illness, and it can promote the idea that a pill is an appropriate response to a social problem, or a modest viral illness.

There are also more concrete harms, specific to the culture in which the placebo is given, rather than the sugar pill itself. For example, it's routine marketing practice for homeopaths to denigrate mainstream medicine. There's a simple commercial reason for this: survey data shows that a disappointing experience with mainstream medicine is almost the only factor that regularly correlates with choosing alternative therapies. This is not just talking medicine down: one study found that more than half of all the homeopaths approached advised patients against the MMR vaccine for their children, acting irresponsibly on what will quite probably come to be known as the media's MMR hoax. How did the alternative therapy world deal with this concerning finding, that so many among them were quietly undermining the vaccination schedule? Prince Charles's office tried to have the lead researcher into the matter sacked.

A BBC *Newsnight* investigation found that almost all the homeopaths approached recommended ineffective homeopathic pills to protect against malaria, and advised against medical malaria prophylactics, while not even giving basic advice on mosquito-bite prevention. This may strike you as neither holistic nor 'complementary'. How did the self-proclaimed 'regulatory bodies' in homeopathy deal with this? None took any action against the homeopaths concerned.

And at the extreme, when they're not undermining public-health campaigns and leaving their patients exposed to fatal diseases, homeopaths who are not medically qualified can miss fatal diagnoses, or actively disregard them, telling their patients grandly to stop using their inhalers, and to throw away their heart pills. There are plenty of examples, but I have too much style to document them here. Suffice to say that while there may be a role for an ethical placebo, homeopaths, at least, have ably demonstrated that they have neither the maturity nor the professionalism to provide it. Fashionable doctors, meanwhile, stunned by the commercial appeal of sugar pills, sometimes wonder – rather unimaginatively – whether they should simply get in on the act and sell some themselves. A smarter idea by far, surely, is to exploit the research we have seen, but only to enhance treatments which really *do* perform better than placebo, and improve healthcare without misleading our patients.

6

The Nonsense *du Jour*

Now we need to raise our game. Food has become, without question, a national obsession. The *Daily Mail* in particular has become engaged in a bizarre ongoing ontological project, diligently sifting through all the inanimate objects of the universe in order to categorise them as a cause of – or cure for – cancer. At the core of this whole project are a small number of repeated canards, basic misunderstandings of evidence which recur with phenomenal frequency.

Although many of these crimes are also committed by journalists, we will be reviewing them later. For the moment we will focus on 'nutritionists', members of a newly invented profession who must create a commercial space to justify their own existence. In order to do this, they must mystify and overcomplicate diet, and foster your dependence upon them. Their profession is based on a set of very simple mistakes in how we interpret scientific literature: they extrapolate wildly from 'laboratory bench data' to make claims about humans; they extrapolate from 'observational data' to make 'intervention claims'; they 'cherry-pick'; and, lastly, they quote published scientific research evidence which seems, as far as one can tell, not to exist.

It's worth going through these misrepresentations of evidence, mainly because they are fascinating illustrations of how people can get things wrong, but also because the aim of this book is that you should be future-proofed against new variants of bullshit. There are also two things we should be very clear on. Firstly, I'm picking out individual examples as props, but these are characteristic of the genre; I could have used many more. Nobody is being bullied, and none of them should be imagined to stand out from the nutritionist crowd, although I'm sure some of the people covered here won't be able to understand how they've done anything wrong.

Secondly, I am not deriding simple, sensible, healthy eating advice. A straightforwardly healthy diet, along with many other aspects of lifestyle (many of which are probably more important, not that you'd know it from reading the papers) is very important. But the media nutritionists speak beyond the evidence: often it is about selling pills; sometimes it is about selling dietary fads, or new diagnoses, or fostering dependence; but it is always driven by their desire to create a market for themselves, in which they are the expert, whereas you are merely bamboozled and ignorant.

Prepare to switch roles.

The four key errors

Does the data exist?

This is perhaps the simplest canard of all, and it happens with surprising frequency, in some rather authoritative venues. Here is Michael van Straten on BBC *Newsnight*, talking 'fact'. If you prefer not to take it on faith that his delivery is earnest, definitive, and perhaps even slightly patrician, you can watch the clip online.

'When Michael van Straten started writing about the magical medicinal powers of fruit juices, he was considered a crank,' *Newsnight* begins. 'But now he finds he's at the forefront of fashion.' (In a world where journalists seem to struggle with science, we should note that *Newsnight* has 'crank' at one end of the axis, and 'fashion' at the other. But that chapter comes later.) Van Straten hands the reporter a glass of juice. 'Two years added to your life expectancy in that!' he chuckles – then a moment of seriousness: 'Well, six months, being honest about it.' A correction. 'A recent study just published last week in America showed that eating pomegranates, pomegranate juice, can actually protect you against ageing, against wrinkles,' he says.

Hearing this on *Newsnight*, the viewer might naturally conclude that a study has recently been published in America showing that pomegranates can protect against ageing. But if you go to Medline, the standard search tool for finding medical academic papers, no such study exists, or at least not that I can find. Perhaps there's some kind of leaflet from the pomegranate industry doing the rounds. He goes on: 'There's a whole group of plastic surgeons in the States who've done a study giving some women pomegranates to eat, and juice to drink, after plastic surgery and before plastic surgery: and they heal in half the time, with half the complications, and no visible wrinkles!' Again, it's a very specific claim – a human trial on pomegranates and surgery – and again, there is nothing in the studies database.

So could you fairly characterise this *Newsnight* performance as 'lying'? Absolutely not. In defence of almost all nutritionists, I would argue that they lack the academic experience, the ill-will, and perhaps even the intellectual horsepower necessary to be fairly derided as liars. The philosopher Professor Harry Frankfurt of Princeton University discusses this issue at length in his classic 1986 essay 'On Bullshit'. Under his model, 'bullshit' is a form of falsehood distinct from lying: the liar knows and cares about the truth, but deliberately sets out to mislead; the truth-speaker knows the truth and is trying to give it to us; the bullshitter, meanwhile, does not care about the truth, and is simply trying to impress us:

It is impossible for someone to lie unless he thinks he knows the truth. Producing bullshit requires no such conviction … When an honest man speaks, he says only what he believes to be true; and for the liar, it is correspondingly indispensable that he considers his statements to be false. For the bullshitter, however, all these bets are off: he is neither on the side of the true nor on the side of the false. His eye is not on the facts at all, as the eyes of the honest man and of the liar are, except insofar as they may be pertinent to his interest in getting away with what he says. He does not care whether the things he says describe reality correctly. He just picks them out, or makes them up, to suit his purpose.

I see van Straten, like many of the subjects in this book, very much in the 'bullshitting' camp. Is it unfair for me to pick out this one man? Perhaps. In biology fieldwork, you throw a wired square called a 'quadrat' at random out onto the ground, and then examine whatever species fall underneath it. This is the approach I have taken with nutritionists, and until I have a Department of Pseudoscience Studies with an army of PhD students doing quantitative work on who is the worst, we shall never know. Van Straten seems like a nice, friendly guy. But we have to start somewhere.

Observation, or intervention?

Does the cock's crow cause the sun to rise? No. Does this light switch make the room get brighter? Yes. Things can happen at roughly the same time, but that is weak, circumstantial evidence for causation. Yet it's exactly this kind of evidence that is used by media nutritionists as confident proof of their claims in our second major canard.

According to the *Daily Mirror*, Angela Dowden, RNutr, is 'Britain's Leading Nutritionist', a moniker it continues to use even though she has been censured by the Nutrition Society for making a claim in the media with literally no evidence whatsoever. Here is a different and more interesting example from Dowden: a quote from her column in the *Mirror*, writing about foods offering protection from the sun during a heatwave: 'An Australian study in 2001 found that olive oil (in combination with fruit, vegetables and pulses) offered measurable protection against skin wrinkling. Eat more olive oil by using it in salad dressings or dip bread in it rather than using butter.'

That's very specific advice, with a very specific claim, quoting a very specific reference, and with a very authoritative tone. It's typical of what you get in the papers from media nutritionists. Let's go to the library and fetch out the paper she refers to ('Skin wrinkling: can food make a difference?' Purba MB et al. *J Am Coll Nutr.* 2001 Feb; 20(1): 71–80). Before we go any further, we should be clear that we are criticising Dowden's *interpretation* of this research, and not the research itself, which we assume is a faithful description of the investigative work that was done.

This was an observational study, not an intervention study. It did not give people olive oil for a time and then measure differences in wrinkles: quite the opposite, in fact. It pooled four different groups of people to get a range of diverse lifestyles, including Greeks, Anglo-Celtic Australians and Swedish people, and it found that people who had completely different eating habits – and completely different lives, we might reasonably assume – also had different amounts of wrinkles.

To me this is not a great surprise, and it illustrates a very simple issue in epidemiological research called 'confounding variables': these are things which are related both to the outcome you're measuring (wrinkles) and to the exposure you are measuring (food), but which you haven't thought of yet. They can confuse an apparently causal relationship, and you have to think of ways to exclude or minimise confounding variables to get to the right answer, or at least be very wary that they are there. In the case of this study, there are almost too many confounding variables to describe.

I eat well – with lots of olive oil, as it happens – and I don't have many wrinkles. I also have a middle-class background, plenty of money, an indoor job, and, if we discount infantile threats of litigation and violence from people who cannot tolerate any discussion of their ideas, a life largely free from strife. People with completely different lives will always have different diets, and different wrinkles. They will have different employment histories, different amounts of stress, different amounts of sun exposure, different levels of affluence, different levels of social support, different patterns of cosmetics use, and much more. I can imagine plenty of reasons why you might find that people who eat olive oil have fewer wrinkles; and the olive oil having a causative role, an actual physical effect on your skin when you eat it, is fairly low down on my list.

Now, to be fair to nutritionists, they are not alone in failing to understand the importance of confounding variables, in their eagerness for a clear story. Every time you read in a newspaper that 'moderate alcohol intake' is associated with some improved health outcome – less heart disease, less obesity, anything – to gales of delight from the alcohol industry, and of course from your friends, who say, 'Ooh well, you see, it's better for me to drink a little …' as they drink a lot – you are almost certainly witnessing a journalist of limited intellect, overinterpreting a study with huge confounding variables.

This is because, let's be honest here: teetotallers are abnormal. They're not like everyone else. They will almost certainly have a reason for not drinking, and it might be moral, or cultural, or perhaps even medical, but there's a serious risk that whatever is causing them to be teetotal might also have other effects on their health, confusing the relationship between their drinking habits and their health outcomes. Like what? Well, perhaps people from specific ethnic groups who are teetotal are also more likely to be obese, so they are less healthy. Perhaps people who deny themselves the indulgence of alcohol are more likely to indulge in chocolate and chips. Perhaps pre-existing ill health will force you to give up alcohol, and that's skewing the figures, making teetotallers look unhealthier than moderate drinkers. Perhaps these teetotallers are recovering alcoholics: among the people I know, they're the ones who are most likely to be absolute teetotallers, and they're also more likely to be fat, from all those years of heavy alcohol abuse. Perhaps some of the people who say they are teetotal are just lying.

This is why we are cautious about interpreting observational data, and to me, Dowden has extrapolated too far from the data, in her eagerness to dispense – with great authority and certainty – *very* specific dietary wisdom in her newspaper column (but of course you may disagree, and you now have the tools to do so meaningfully).

If we were modern about this, and wanted to offer construc-tive criticism, what might she have written instead? I think, both here and elsewhere, that despite what journalists and self-appointed 'experts' might say, people are perfectly capable of understanding the evidence for a claim, and anyone who with-holds, overstates or obscures that evidence, while implying that they're doing the reader a favour, is probably up to no good. MMR is an excellent parallel example of where the bluster, the panic, the 'concerned experts' and the conspiracy theories of the media were very compelling, but the science itself was rarely explained.

So, leading by example, if I were a media nutritionist, I might say, if pushed, after giving all the other sensible sun advice: 'A survey found that people who eat more olive oil have fewer wrinkles,' and I might feel obliged to add, 'Although people with different diets may differ in lots of other ways.' But then, I'd also be writing about food, so: 'Never mind, here's a delicious recipe for salad dressing anyway.' Nobody's going to employ me to write a nutritionist column.

From the lab bench to the glossies

Nutritionists love to quote basic laboratory science research because it makes them look as if they are actively engaged in a process of complicated, impenetrable, highly technical academic work. But you have to be very cautious about how you extrapolate from what happens to some cells in a dish, on a laboratory bench, to the complex system of a living human being, where things can work in completely the opposite way to what laboratory work would suggest. Anything can kill cells in a test tube. Fairy Liquid will kill cells in a test tube, but you don't take it to cure cancer. This is just another example of how nutritionism, despite the 'alternative medicine' rhetoric and phrases like 'holistic', is actually a crude, unsophisticated, old fashioned, and above all *reductionist* tradition.

Later we will see Patrick Holford, the founder of the Institute for Optimum Nutrition, stating that vitamin C is better than the AIDS drug AZT on the basis of an experiment where vitamin C was tipped onto some cells in a dish. Until then, here is an example from Michael van Straten – who has fallen sadly into our quadrat, and I don't want to introduce too many new characters or confuse you – writing in the *Daily Express* as its nutrition specialist: 'Recent research', he says, has shown that turmeric is 'highly protective against many forms of cancer, especially of the prostate'. It's an interesting idea, worth pursuing, and there have been some speculative lab studies of cells, usually from rats, growing or not growing under microscopes, with turmeric extract tipped on them. There is some limited animal model data, but it is not fair to say that turmeric, or curry, in the real world, in real people, is 'highly protective against many forms of cancer, especially of the prostate', least of all because it's not very well absorbed.

Forty years ago a man called Austin Bradford-Hill, the grandfather of modern medical research, who was key in discovering the link between smoking and lung cancer, wrote out a set of guidelines, a kind of tick list, for assessing causality, and a relationship between an exposure and an outcome. These are the cornerstone of evidence-based medicine, and often worth having at the back of your mind: it needs to be a strong association, which is consistent, and specific to the thing you are studying, where the putative cause comes before the supposed effect in time; ideally there should be a biological gradient, such as a dose–response effect; it should be consistent, or at least not completely at odds with, what is already known (because extraordinary claims require extraordinary evidence); and it should be biologically plausible.

Michael van Straten, here, has got biological plausibility, and little else. Medics and academics are very wary of people making claims on such tenuous grounds, because it's something you get a lot from people with something to sell: specifically, drug companies. The public don't generally have to deal with drug-company propaganda, because the companies are not currently allowed to talk to patients in Europe – thankfully – but they badger doctors incessantly, and they use many of the same tricks as the miracle-cure industries. You're taught about these tricks at medical school, which is how I'm able to teach you about them now.

Drug companies are very keen to promote theoretical advantages ('It works more on the Z4 receptor, so it must have fewer side-effects!'), animal experiment data or 'surrogate outcomes' ('It improves blood test results, it must be protective against heart attacks!') as evidence of the efficacy or superiority of their product. Many of the more detailed popular nutritionist books, should you ever be lucky enough to read them, play this classic drug-company card very assertively. They will claim, for example, that a 'placebo-controlled randomised control trial' has shown *benefits* from a particular vitamin, when what they mean is, it showed changes in a 'surrogate outcome'.

For example, the trial may merely have shown that there were measurably increased amounts of the vitamin in the bloodstream after taking a vitamin, compared to placebo, which is a pretty unspectacular finding in itself: yet this is presented to the unsuspecting lay reader as a positive trial. Or the trial may have shown that there were changes in some other blood marker, perhaps the level of an ill-understood immune-system component, which, again, the media nutritionist will present as concrete evidence of a real-world benefit.

There are problems with using such surrogate outcomes. They are often only tenuously associated with the real disease, in a very abstract theoretical model, and often developed in the very idealised world of an experimental animal, genetically inbred, kept under conditions of tight physiological control. A surrogate outcome can – of course – be used to generate and examine hypotheses about a real disease in a real person, but it needs to be very carefully validated. Does it show a clear dose–response relationship? Is it a true predictor of disease, or merely a 'co-variable', something that is related to the disease in a different way (e.g. caused *by* it rather than involved in *causing* it)? Is there a well-defined cut-off between normal and abnormal values?

All I am doing, I should be clear, is taking the fêted media nutritionists at their own word: they present themselves as men and women of science, fill their columns, TV shows and books with references to scientific research. I am subjecting their claims to the exact same level of very basic, uncomplicated rigour that I would deploy for any new theoretical work, any drug company claim and pill marketing rhetoric, and so on.

It's not unreasonable to use surrogate outcome data, as they do, but those who are in the know are always circumspect. We're *interested* in early theoretical work, but often the message is: 'It might be a bit more complicated than that …'. You'd only want to accord a surrogate outcome any significance if you'd read everything on it yourself, or if you could be absolutely certain that the person assuring you of its validity was extremely capable, and was giving a sound appraisal of all the research in a given field, and so on.

Similar problems arise with animal data. Nobody could deny that this kind of data is valuable in the theoretical domain, for developing hypotheses, or suggesting safety risks, when cautiously appraised. But media nutritionists, in their eagerness to make lifestyle claims, are all too often blind to the problems of applying these isolated theoretical nuggets to humans, and anyone would think they were just trawling the internet looking for random bits of science to sell their pills and expertise (imagine that). Both the tissue and the disease in an animal model, after all, may be very different to those in a living human system, and these problems are even greater with a lab-dish model. Giving unusually high doses of chemicals to animals can distort the usual metabolic pathways, and give misleading results – and so on. Just because something can upregulate or downregulate something in a model doesn't mean it will have the effect you expect in a person – as we will see with the stunning truth about antioxidants.

And what about turmeric, which we were talking about before I tried to show you the entire world of applying theoretical research in this tiny grain of spice? Well, yes, there is some evidence that curcumin, a chemical in turmeric, is highly biologically active, in all kinds of different ways, on all kinds of different systems (there are also theoretical grounds for believing that it may be carcinogenic, mind you). It's certainly a valid target for research.

But for the claim that we should eat more curry in order to get more of it, that 'recent research' has shown it is 'highly protective against many forms of cancer, especially of the prostate', you might want to step back and put the theoretical claims in the context of your body. Very little of the curcumin you eat is absorbed. You have to eat a few grams of it to reach significant detectable serum levels, but to get a few grams of *curcumin,* you'd have to eat 100g of *turmeric:* and good luck with that. Between research and recipe, there's a lot more to think about than the nutritionists might tell you.

Cherry-picking

The idea is to try and give all the information to help others to judge the value of your contribution; not just the information that leads to judgment in one particular direction or another.

Richard P. Feynman

There have been an estimated fifteen million medical academic articles published so far, and 5,000 journals are published every month. Many of these articles will contain contradictory claims: picking out what's relevant – and what's not – is a gargantuan task. Inevitably people will take shortcuts. We rely on review articles, or on meta-analyses, or textbooks, or hearsay, or chatty journalistic reviews of a subject.

That's if your interest is in getting to the truth of the matter. What if you've just got a point to prove? There are few opinions so absurd that you couldn't find at least one person with a PhD somewhere in the world to endorse them for you; and similarly, there are few propositions in medicine so ridiculous that you couldn't conjure up some kind of published experimental evidence somewhere to support them, if you didn't mind it being a tenuous relationship, and cherry-picked the literature, quoting only the studies that were in your favour.

One of the great studies of cherry-picking in the academic literature comes from an article about Linus Pauling, the great-grandfather of modern nutritionism, and his seminal work on vitamin C and the common cold. In 1993 Paul Knipschild,

Professor of Epidemiology at the University of Maastricht, published a chapter in the mighty textbook *Systematic Reviews*: he had gone to the extraordinary trouble of approaching the literature as it stood when Pauling was working, and subjecting it to the same rigorous systematic review that you would find in a modern paper.

He found that while some trials did suggest that vitamin C had some benefits, Pauling had selectively quoted from the literature to prove his point. Where Pauling had referred to some trials which seriously challenged his theory, it was to dismiss them as methodologically flawed: but as a cold examination showed, so too were papers he quoted favourably in support of his own case.

In Pauling's defence, his was an era when people knew no better, and he was probably quite unaware of what he was doing: but today cherry-picking is one of the most common dubious practices in alternative therapies, and particularly in nutritionism, where it seems to be accepted essentially as normal practice (it is this cherry-picking, in reality, which helps to characterise what alternative therapists conceive of rather grandly as their 'alternative paradigm'). It happens in mainstream medicine also, but with one crucial difference: there it is recognised as a major problem, and hard work has been done to derive a solution.

That solution is a process called 'systematic review'. Instead of just mooching around online and picking out your favourite papers to back up your prejudices and help you sell a product, in a systematic review you have an explicit search strategy for seeking out data (openly described in your paper, even including the search terms you used on databases of research papers), you tabulate the characteristics of each study you find, you measure – ideally blind to the results – the methodological quality of each one (to see how much of a 'fair test' it is), you compare alternatives, and then finally you give a critical, weighted summary.

This is what the Cochrane Collaboration does on all the healthcare topics that it can find. It even invites people to submit new clinical questions that need an answer. This careful sifting of information has revealed huge gaps in knowledge, it has revealed that 'best practices' were sometimes murderously flawed, and simply by sifting methodically through pre-existing data, it has saved more lives than you could possibly imagine. In the nineteenth century, as the public-health doctor Muir Gray has said, we made great advances through the provision of clean, clear water; in the twenty-first century we will make the same advances through clean, clear information. Systematic reviews are one of the great ideas of modern thought. They should be celebrated.

Problematising antioxidants

We have seen the kinds of errors made by those in the nutritionism movement as they strive to justify their more obscure and technical claims. What's more fun is to take our new understanding and apply it to one of the key claims of the nutritionism movement, and indeed to a fairly widespread belief in general: the claim that you should eat more antioxidants.

As you now know, there are lots of ways of deciding whether the totality of research evidence for a given claim stacks up, and it's rare that one single piece of information clinches it. In the case of a claim about food, for example, there are all kinds of different things we might look for: whether it is theoretically plausible, whether it is backed up by what we know from observing diets and health, whether it is supported by 'intervention trials' where we give one group one diet and another group a different one, and whether those trials measured real-world outcomes, like 'death', or a surrogate outcome, like a blood test, which is only hypothetically related to a disease.

My aim here is by no means to suggest that antioxidants are *entirely* irrelevant to health. If I had a T-shirt slogan for this whole book it would be: 'I think you'll find it's a bit more complicated than that'. I intend, as they say, to 'problematise' the prevailing nutritionist view on antioxidants, which currently lags only about twenty years behind the research evidence.

From an entirely theoretical perspective, the idea that antioxidants are beneficial for health is an attractive one. When I was a medical student – not so long ago – the most popular biochemistry textbook was called Stryer. This enormous book is filled with complex interlocking flow charts of how chemicals – which is what you are made of – move through the body. It shows how different enzymes break down food into its constituent molecular elements, how these are absorbed, how they are reassembled into new larger molecules that your body needs to build muscles, retina, nerves, bone, hair, membrane, mucus, and everything else that you're made of; how the various forms of fats are broken down, and reassembled into new forms of fat; or how different forms of molecule – sugar, fat, even alcohol – are broken down gradually, step by step, to release energy, and how that energy is transported, and how the incidental products from that process are used, or bolted onto something else to be transported in the blood, and then ditched at the kidneys, or metabolised down into further constituents, or turned into something useful elsewhere, and so on. This is one of the great miracles of life, and it is endlessly, beautifully, intricately fascinating.

Looking at these enormous, overwhelming interlocking webs, it's hard not to be struck by the versatility of the human body, and how it can perform acts of near alchemy from so many different starting points. It would be very easy to pick one of the elements of these vast interlocking systems and become fixated on the idea that it is uniquely important. Perhaps it appears a lot on the diagram; or perhaps rarely, and seems to serve a uniquely important function in one key place. It would be easy to assume that if there was more of it around, then that function would be performed with greater efficiency.

But, as with all enormous interlocking systems – like societies, for example, or businesses – an intervention in one place can have quite unexpected consequences: there are feedback mechanisms, compensatory mechanisms. Rates of change in one localised area can be limited by quite unexpected factors that are entirely remote from what you are altering, and excesses of one thing in one place can distort the usual pathways and flows, to give counterintuitive results.

The theory underlying the view that antioxidants are good for you is the 'free radical theory of ageing'. Free radicals are highly chemically reactive, as are many things in the body. Often this reactivity is put to very good use. For example, if you have an infection, and there are some harmful bacteria in your body, then a phagocytic cell from your immune system might come along, identify the bacteria as unwelcome, build a strong wall around as many of them as it can find, and blast them with destructive free radicals. Free radicals are basically like bleach, and this process is a lot like pouring bleach down the toilet. Once again, the human body is cleverer than anybody you know.

But free radicals in the wrong places can damage the desirable components of cells. They can damage the lining of your arteries, and they can damage DNA; and damaged DNA leads to ageing or cancer, and so on. For this reason, it has been suggested that free radicals are responsible for ageing and various diseases. This is a theory, and it may or may not be correct.

Antioxidants are compounds which can – and do – 'mop up' these free radicals, by reacting with them. If you look at the vast, interlocking flow chart diagrams of how all the molecules in your body are metabolised from one form to the next, you can see that this is happening all over the shop.

The theory that antioxidants are protective is separate to – but builds upon – the free radical theory of disease. If free radicals are dangerous, the argument goes, and antioxidants on the big diagrams are involved in neutralising them, then eating more antioxidants should be good for you, and reverse or slow ageing, and prevent disease.

There are a number of problems with this as a theory. Firstly, who says free radicals are always bad? If you're going to reason just from theory, and from the diagrams, then you can hook all kinds of things together and make it seem as if you're talking sense. As I said, free radicals are vital for your body to kill off bacteria in phagocytic immune cells: so should you set yourself up in business and market an antioxidant-*free* diet for people with bacterial infections?

Secondly, just because antioxidants are involved in doing something good, why should eating more of them necessarily make that process more efficient? I know it makes sense superficially, but so do a lot of things, and that's what's really interesting about science (and this story in particular): sometimes the results aren't quite what you might expect. Perhaps an excess of antioxidants is simply excreted, or turned into something else. Perhaps it just sits there doing nothing, because it's not needed. After all, half a tank of petrol will get you across town just as easily as a full tank. Or perhaps, if you have an unusually enormous amount of antioxidant lying around in your body doing nothing, it doesn't just do nothing. Perhaps it does something actively harmful. That would be a turn-up for the books, wouldn't it?

There were a couple of other reasons why the antioxidant theory seemed like a good idea twenty years ago. Firstly, when you take a static picture of society, people who eat lots of fresh fruit and vegetables tend to live longer, and have less cancer and heart disease; and there are lots of antioxidants in fruit and vegetables (although there are lots of other things in them too, and, you might rightly assume, lots of other healthy things about the lives of people who eat lots healthy fresh fruit and vegetables, like their posh jobs, moderate alcohol intake, etc.).

Similarly, when you take a snapshot picture of the people who take antioxidant supplement pills, you will often find that they are healthier, or live longer: but again (although nutritionists are keen to ignore this fact), these are simply surveys of people who have already chosen to take vitamin pills. These are people who are more likely to care about their health, and are different from the everyday population – and perhaps from you – in lots of other ways, far beyond their vitamin pill consumption: they may take more exercise, have more social supports, smoke less, drink less, and so on.

But the early evidence in favour of antioxidants was genuinely promising, and went beyond mere observational data on nutrition and health: there were also some very seductive blood results. In 1981 Richard Peto, one of the most famous epidemiologists in the world, who shares the credit for discovering that smoking causes 95 per cent of lung cancer, published a major paper in *Nature*. He reviewed a number of studies which apparently showed a positive relationship between having a lot of β-carotene onboard (this is an antioxidant available in the diet) and a reduced risk of cancer.

This evidence included 'case-control studies', where people *with* various cancers were compared against people *without* cancer (but matched for age, social class, gender and so on), and it was found that the cancer-free subjects had higher plasma carotene. There were also 'prospective cohort studies', in which people were classified by their plasma carotene level at the beginning of the study, before any of them had cancer, and then followed up for many years. These studies showed twice as much lung cancer in the group with the lowest plasma carotene, compared with those with the highest level. It looked as if having more of these antioxidants might be a very good thing.

Similar studies showed that higher plasma levels of antioxidant vitamin E were related to lower levels of heart disease. It was suggested that vitamin E status explained much of the variations in levels of ischaemic heart disease between different countries in Europe, which could not be explained by differences in plasma cholesterol or blood pressure.

But the editor of *Nature* was cautious. A footnote was put onto the Peto paper which read as follows:

> Unwary readers (if such there are) should not take the accompanying article as a sign that the consumption of large quantities of carrots (or other dietary sources of β-carotene) is necessarily protective against cancer.

It was a very prescient footnote indeed.

The antioxidant dream unravels

Whatever the shrill alternative therapists may say, doctors and academics have an interest in chasing hints that could bear fruit, and compelling hypotheses like these – which could save millions of lives – are not taken lightly. These studies were acted upon, with many huge trials of vitamins set up and run around the world. There's also an important cultural context for this rush of activity which cannot be ignored: it was the tail end of the golden age of medicine. Before 1935 there weren't too many effective treatments around: we had insulin, liver for iron deficiency anaemia, and morphine – a drug with superficial charm at least – but in many respects, doctors were fairly useless. Then suddenly, between about 1935 and 1975, science poured out a constant stream of miracles.

Almost everything we associate with modern medicine happened in that time: treatments like antibiotics, dialysis, transplants, intensive care, heart surgery, almost every drug you've ever heard of, and more. As well as the miracle treatments, we really were finding those simple, direct, hidden killers that the media still pine for so desperately in their headlines. Smoking, to everybody's genuine surprise – one single risk factor – turned out to cause almost all lung cancer. And asbestos, through some genuinely brave and subversive investigative work, was shown to cause mesothelioma.

The epidemiologists of the 1980s were on a roll, and they believed that they were going to find lifestyle causes for all the major diseases of humankind. A discipline that had got cracking when John Snow took the handle off the Broad Street pump in 1854, terminating that pocket of the Soho cholera epidemic by cutting off the supply of contaminated water (it was a bit more complicated than that, but we don't have the time here) was going to come into its own. They were going to identify more and more of these one-to-one correlations between exposures and disease, and, in their fervent imaginations, with simple interventions and cautionary advice they were going to save whole nations of people. This dream was very much not realised, as it turned out to be a bit more complicated than that.

Two large trials of antioxidants were set up after Peto's paper (which rather gives the lie to nutritionists' claims that vitamins are never studied because they cannot be patented: in fact there have been a great many such trials, although the food supplement industry, estimated by one report to be worth over $50 billion globally, rarely deigns to fund them). One was in Finland, where 30,000 participants at high risk of lung cancer were recruited, and randomised to receive either β-carotene, vitamin E, or both, or neither. Not only was there an increase in death from lung cancer among the people receiving the supposedly protective β-carotene supplements, compared with placebo, but the vitamin group also showed increased death from prostate and gastric cancers.

The results of the other trial were almost worse. It was called the 'Carotene and Retinol Efficacy Trial', or 'CARET', in honour of the high β-carotene content of carrots. It's interesting to note, while we're here, that carrots were the source of one of the great disinformation coups of World War II, when the Germans couldn't understand how our pilots could see their planes coming from huge distances, even in the dark. To stop them trying to work out if we'd invented anything clever like radar (which we had), the British instead started an elaborate and entirely made-up nutritionist rumour. Carotenes in carrots, they explained, are transported to the eye and converted to retinal, which is the molecule that detects light in the eye (this is basically true, and is a plausible mechanism, like those we've already dealt with): so, went the story, doubtless with much chortling behind their excellent RAF moustaches, we have been feeding our chaps huge plates of carrots, to jolly good effect.

Anyway. Two groups of people at high risk of lung cancer were studied: smokers, and people who had been exposed to asbestos at work. Half were given β-carotene and vitamin A, while the other half got placebo. Eighteen thousand participants were due to be recruited throughout its course, and the intention was that they would be followed up for an average of six years; but in fact the trial was terminated early, because it was considered unethical to continue it. Why? The people having the antioxidant tablets were 46 per cent more likely to die from lung cancer, and 17 per cent more likely to die of any cause,* than the people taking placebo pills. This is not news, hot off the presses: it happened well over a decade ago.

Since then the placebo-controlled trial data on antioxidant vitamin supplements has continued to give negative results. The most up-to-date Cochrane reviews of the literature pool together all the trials on the subject, after sourcing the widest

---

* I have deliberately expressed this risk in terms of the 'relative risk increase', as part of a dubious in-joke with myself. You will learn about this on page 240.

possible range of data using the systematic search strategies described above (rather than 'cherry-picking' studies to an agenda): they assess the quality of the studies, and then put them all into one giant spreadsheet to give the most accurate possible estimate of the risks of benefits, and they show that antioxidant supplements are either ineffective, or perhaps even actively harmful.

The Cochrane review on preventing lung cancer pooled data from four trials, describing the experiences of over 100,000 participants, and found no benefit from antioxidants, and indeed an increase in risk of lung cancer in participants taking β-carotene and retinol together. The most up-to-date systematic review and meta-analysis on the use of antioxidants to reduce heart attacks and stroke looked at vitamin E, and separately β-carotene, in fifteen trials, and found no benefit for either. For β-carotene, there was a small but significant increase in death.

Most recently, a Cochrane review looked at the number of deaths, from any cause, in all the placebo-controlled randomised trials on antioxidants which have ever been performed (many of which looked at quite high doses, but perfectly in line with what you can buy in health-food stores), describing the experiences of 230,000 people in total. This showed that overall, antioxidant vitamin pills do not reduce deaths, and in fact they may increase your chance of dying.

Where does all this leave us? There was an observed correlation between low blood levels of these antioxidant nutrients and a higher incidence of cancer and heart disease, and a plausible mechanism for how they could have been preventive: but when you gave them as supplements, it turned out that people were no better off, or were possibly *more* likely to die. That is, in some respects, a shame, as nice quick fixes are always useful, but there you go. It means that something funny is going on, and it will be interesting to get to the bottom of it and find out what.

More interesting is how uncommon it is for people even to be aware of these findings about antioxidants. There are various reasons why this has happened. Firstly, it's an unexpected finding, although in that regard antioxidants are hardly an isolated case. Things that work in theory often do not work in practice, and in such cases we need to revise our theories, even if it is painful. Hormone replacement therapy seemed like a good idea for many decades, until the follow-up studies revealed the problems with it, so we changed our views. And calcium supplements once looked like a good idea for osteo-porosis, but now it turns out that they probably increase the risk of heart attacks in older women, so we change our view.

It's a chilling thought that when we think we are doing good, we may actually be doing harm, but it is one we must always be alive to, even in the most innocuous situations. The paediatri-cian Dr Benjamin Spock wrote a record-breaking best-seller called *Baby and Child Care*, first published in 1946, which was hugely influential and largely sensible. In it, he confidently recommended that babies should sleep on their tummies. Dr Spock had little to go on, but we now know that this advice is wrong, and the apparently trivial suggestion contained in his book, which was so widely read and followed, has led to thou-sands, and perhaps even tens of thousands, of avoidable cot deaths. The more people are listening to you, the greater the effects of a small error can be. I find this simple anecdote deeply disturbing.

But of course, there is a more mundane reason why people may not be aware of these findings on antioxidants, or at least may not take them seriously, and that is the phenomenal lobby-ing power of a large, sometimes rather dirty industry, which sells a lifestyle product that many people feel passionately about. The food supplement industry has engineered itself a beneficent public image, but this is not borne out by the facts. Firstly, there is essentially no difference between the vitamin industry and the pharmaceutical and biotech industries (that is one message of this book, after all: the tricks of the trade are the same the world over). Key players include companies like Roche and Aventis; BioCare, the vitamin pill company that media nutritionist Patrick Holford works for, is part-owned by Elder Pharmaceuticals, and so on. The vitamin industry is also – amusingly – legendary in the world of economics as the setting of the most outrageous price-fixing cartel ever documented. During the 1990s the main offenders were forced to pay *the largest criminal fines ever levied in legal history* – $1.5 billion in total – after entering guilty pleas with the US Department of Justice and regulators in Canada, Australia and the European Union. That's quite some cosy cottage industry.

Whenever a piece of evidence is published suggesting that the $50-billion food supplement pill industry's products are ineffective, or even harmful, an enormous marketing machine lumbers into life, producing spurious and groundless methodological criticisms of the published data in order to muddy the waters – not enough to be noteworthy in a meaningful academic discussion, but that is not their purpose. This is a well-worn risk-management tactic from many industries, including those producing tobacco, asbestos, lead, vinyl chloride, chromium and more. It is called 'manufacturing doubt', and in 1969 one tobacco executive was stupid enough to commit it to paper in a memo: 'Doubt is our product,' he wrote, 'since it is the best means of competing with the "body of fact" that exists in the minds of the general public. It is also the means of establishing a controversy.'

Nobody in the media dares to challenge these tactics, where lobbyists raise sciencey-sounding defences of their products, because they feel intimidated, and lack the skills to do so. Even if they did, there would simply be a confusing and technical discussion on the radio, which everyone would switch off, and at most the consumer would hear only 'controversy': job done.

I don't think that food supplement pills are as dangerous as tobacco – few things are – but it's hard to think of any other kind of pill where research could be published showing a possible increase in death, and industry figures would be wheeled out and given as easy a ride as the vitamin companies' employees are given when papers are published on their risks. But then, of course, many of them have their own slots in the media to sell their wares and their world view.

The antioxidant story is an excellent example of how wary we should be of blindly following hunches based on laboratory-level and theoretical data, and naïvely assuming, in a reductionist manner, that this must automatically map onto dietary and supplement advice, as the media nutritionists would have us do. It is an object lesson in what an unreliable source of research information these characters can be, and we would all do well to remember this story the next time someone tries to persuade us with blood test data, or talk about molecules, or theories based on vast, interlocking metabolism diagrams, that we should buy their book, their wacky diet, or their bottle of pills.

More than anything it illustrates how this atomised, over-complicated view of diet can be used to mislead and oversell. I don't think it's melodramatic to speak of people disempowered and paralysed by confusion, with all the unnecessarily complex and conflicting messages about food. If you're really worried, you can buy Fruitella Plus with added vitamins A, C, E and calcium, and during Christmas 2007 two new antioxidant products came on the market, the ultimate expression of how nutritionism has perverted and distorted our common sense about food. Choxi+ is milk chocolate with 'extra antioxidants'. The *Daily Mirror* says it's 'too good to be true'. It's 'chocolate that is good for you, as well as seductive', according to the *Daily Telegraph*. 'Guilt free', says the *Daily Mail*: it's 'the chocolate bar that's "healthier" than 5lb of apples'. The company even 'recommends' two pieces of its chocolate a day. Meanwhile, Sainsbury's is promoting Red Heart wine – with extra antioxidants – as if drinking the stuff was a duty to your grandchildren.

If I was writing a lifestyle book it would have the same advice on every page, and you'd know it all already. Eat lots of fruit and vegetables, and live your whole life in every way as well as you can: exercise regularly as part of your daily routine, avoid obesity, don't drink too much, don't smoke, and don't get distracted from the real, basic, simple causes of ill health. But as we will see, even these things are hard to do on your own, and in reality require wholesale social and political changes.

7

Dr Gillian McKeith PhD

I'm going to push the boat out here, and suggest that since you've bought this book you may already be harbouring some suspicions about multi-millionaire pill entrepreneur and clinical nutritionist Gillian McKeith (or, to give her full medical title: Gillian McKeith).

She is an empire, a prime-time TV celebrity, a best-selling author. She has her own range of foods and mysterious powders, she has pills to give you an erection, and her face is in every health food store in the country. Scottish Conservative politicians want her to advise the government. The Soil Association gave her a prize for educating the public. But to anyone who knows even the slightest bit about science, she is a joke.

The most important thing to recognise is that there is nothing new here. Although the contemporary nutritionism movement likes to present itself as a thoroughly modern and evidence-based enterprise, the food-guru industry, with its outlandish promises, moralising and sexual obsessions, goes back at least two centuries.

Like our modern food gurus, the historical figures of nutritionism were mostly enthusiastic lay people, and they all claimed to understand nutritional science, evidence and medicine better than the scientists and doctors of their era. The advice and the products may have shifted with prevailing religious and moral notions, but they have always played to the market, be it puritan or liberal, New Age or Christian.

Graham crackers are a digestive biscuit invented in the nineteenth century by Sylvester Graham, the first great advocate of vegetarianism and nutritionism as we would know it, and proprietor of the world's first health food shop. Like his descendants today, Graham mixed up sensible notions – such as cutting down on cigarettes and alcohol – with some other, rather more esoteric, ideas which he concocted for himself. He warned that ketchup and mustard, for example, can cause 'insanity'.

I've got no great beef with the organic food movement (even if its claims are a little unrealistic), but it's still interesting to note that Graham's health food store – in 1837 – heavily promoted its food as being grown according to 'physiological principles' on 'virgin unvitiated soil'. By the retro-fetishism of the time, this was soil which had not been 'subjected' to 'over-stimulation' … by manure.

Soon these food marketing techniques were picked up by more overtly puritanical religious zealots like John Harvey Kellogg, the man behind the cornflake. Kellogg was a natural healer, anti-masturbation campaigner, and health food advocate, promoting his granola bars as the route to abstinence, temperance and solid morals. He ran a sanatorium for private clients, using 'holistic' techniques, including Gillian McKeith's favourite, colonic irrigation.

Kellogg was also a keen anti-masturbation campaigner. He advocated exposing the tissue on the end of the penis, so that it smarted with friction during acts of self-pollution (and you do have to wonder about the motives of anyone who thinks the problem through in that much detail). Here is a particularly enjoyable passage from his *Treatment for Self-Abuse and its Effects* (1888), in which Kellogg outlines his views on circumcision:

The operation should be performed by a surgeon without administering an anesthetic, as the brief pain attending the operation will have a salutary effect upon the mind, especially if it be connected with the idea of punishment. In females, the author has found the application of pure carbolic acid to the clitoris an excellent means of allaying the abnormal excitement.

By the early twentieth century a man called Bernard Macfadden had updated the nutritionism model for contemporary moral values, and so became the most commercially successful health guru of his time. He changed his Christian name from Bernard to Bernarr, because it sounded more like the roar of a lion (this is completely true), and ran a successful magazine called *Physical Culture,* featuring beautiful bodies doing healthy things. The pseudoscience and the posturing were the same, but he used liberal sexuality to his advantage, selling his granola bars as a food that would promote a muscular, thrusting, lustful lifestyle in that decadent rush that flooded the populations of the West between the wars.*

More recently there was Dudley J. LeBlanc, a Louisiana Senator and the man behind Hadacol ('I had'da call it some-

---

* Interestingly, Macfadden's food product range was complemented by a more unusual invention of his own. The 'Peniscope' was a popular suction device designed to enlarge the male organ which is still used by many today, in a modestly updated form. Since this may be your only opportunity to learn about the data on penis enlargement, it's worth mentioning that there is, in fact, some evidence that stretching devices can increase penis size. Gillian McKeith's 'Wild Pink' and 'Horny Goat Weed' sex supplement pills, however, sold for 'maintaining erections, orgasmic pleasure, ejaculation … lubrication, satisfaction, and arousal', could claim no such evidence for efficacy (and in 2007, after much complaining, these seedy and rather old-fashioned products were declared illegal by the Medicines and Healthcare Products Regulatory Agency, or MHRA). I mention this only because, rather charmingly, it means that Macfadden's Peniscope may have a better evidence base for its claims than either his own food products or McKeith's Horny Goat penis pills.

thing'). It cured everything, cost $100 a year for the recom-
mended dose, and to Dudley's open amazement, it sold in
millions. 'They came in to buy Hadacol,' said one pharmacist,
'when they didn't have money to buy food. They had holes in
their shoes and they paid $3.50 for a bottle of Hadacol.'

LeBlanc made no medicinal claims, but pushed customer
testimonials to an eager media. He appointed a medical director
who had been convicted in California of practising medicine
with no licence and no medical degree. A diabetic patient
almost died when she gave up insulin to treat herself with
Hadacol, but nobody cared. 'It's a craze. It's a culture. It's a political movement,' said *Newsweek*.

It's easy to underestimate the phenomenal and enduring
commercial appeal of these kinds of products and claims
throughout history. By 1950 Hadacol's sales were over $20
million, with an advertising spend of $1 million a month, in 700
daily papers and on 528 radio stations. LeBlanc took a travelling
medicine show of 130 vehicles on a tour of 3,800 miles through
the South. Entry was paid in Hadacol bottle tops, and the shows
starred Groucho and Chico Marx, Mickey Rooney, Judy
Garland, and educational exhibitions of scantily clad women
illustrating 'the history of the bathing suit'. Dixieland bands
played songs like 'Hadacol Boogie' and 'Who Put the Pep in
Grandma?'.

The Senator used Hadacol's success to drive his political
career, and his competitors, the Longs – descended from the
Democrat reformer Huey Long – panicked, launching their
own patent medicine called 'Vita-Long'. By 1951 LeBlanc was
spending more in advertising that he was making in sales, and
in February of that year, shortly after he sold the company – and
shortly before it folded – he appeared on the TV show *You Bet
Your Life* with his old friend Groucho Marx. 'Hadacol,' said
Groucho, 'what's that good for?' 'Well,' said LeBlanc, 'it was good
for about five and a half million dollars for me last year.'

The point I am making is that there is nothing new under the sun. There have always been health gurus selling magic potions. But I am not a consumer journalist, and I don't care if people have unusual qualifications, or sell silly substances. McKeith is, for me, very simply a menace to the public understanding of science. She has a mainstream prime-time television nutrition show, yet she seems to misunderstand not nuances, but the most basic aspects of biology, things that a schoolchild could put her straight on.

I first noticed Dr Gillian McKeith when a reader sent in a clipping from the *Radio Times* about her first series on Channel 4. McKeith was styled, very strikingly, as a white-coated academic and scientific authority on nutrition, a 'clinical nutritionist', posing in laboratories, surrounded by test tubes, and talking about diagnoses and molecules. She was also quoted here saying something a fourteen-year-old doing GCSE biology could easily have identified as pure nonsense: recommending spinach, and the darker leaves on plants, because they contain more chlorophyll. According to McKeith these are 'high in oxygen' and will 'really oxygenate your blood'. This same claim is repeated all over her books.

Forgive me for patronising, but before we go on you may need a little refresher on the miracle of photosynthesis. Chlorophyll is a small green molecule which is found in chloroplasts, the miniature factories in plant cells that take the energy from sunlight and use it to convert carbon dioxide and water into sugar and oxygen. Using this process, called photosynthesis, plants store the energy from sunlight in the form of sugar (high in calories, as you know), and they can then use this sugar energy to make everything else they need: like protein, and fibre, and flowers, and corn on the cob, and bark, and leaves, and amazing traps that eat flies, and cures for cancer, and tomatoes, and wispy dandelions, and conkers, and chillies, and all the other amazing things that the plant world has going on.

Meanwhile, you breathe in the oxygen that the plants give off during this process – essentially as a byproduct of their sugar manufacturing – and you also eat the plants, or you eat animals that eat the plants, or you build houses out of wood, or you make painkiller from willow bark, or any of the other amazing things that happen with plants. You also breathe out carbon dioxide, and the plants can combine that with water to make more sugar again, using the energy from sunlight, and so the cycle continues.

Like most things in the story the natural sciences can tell about the world, it's all so beautiful, so gracefully simple, yet so rewardingly complex, so neatly connected – not to mention true – that I can't even begin to imagine why anyone would ever want to believe some New Age 'alternative' nonsense instead. I would go so far as to say that even if we are all under the control of a benevolent God, and the whole of reality turns out to be down to some flaky spiritual 'energy' that only alternative therapists can truly harness, that's still neither so interesting nor so graceful as the most basic stuff I was taught at school about how plants work.

Is chlorophyll 'high in oxygen'? No. It helps to make oxygen. In sunlight. And it's pretty dark in your bowels: in fact, if there's any light in there at all then something's gone badly wrong. So any chlorophyll you eat will not create oxygen, and even if it did, even if Dr Gillian McKeith PhD stuck a searchlight right up your bum to prove her point, and your salad began photosynthesising, even if she insufflated your guts with carbon dioxide through a tube, to give the chloroplasts something to work with, and by some miracle you really did start to produce oxygen in there, you still wouldn't absorb a significant amount of it through your bowel, because your bowel is adapted to absorb food, while your lungs are optimised to absorb oxygen. You do not have gills in your bowels. Neither, since we've mentioned them, do fish. And while we're talking about it, you probably don't want oxygen inside your abdomen anyway: in keyhole surgery, surgeons have to inflate your abdomen to help them see what they're doing, but they don't use oxygen, because there's methane fart gas in there too, and we don't want anyone catching fire on the inside. There is no oxygen in your bowel.

So who is this person, and how did she come to be teaching us about diet, on a prime-time television show, on a national terrestrial channel? What possible kind of science degree can she have, to be making such basic mistakes that a schoolkid would spot? Was this an isolated error? A one-off slip of the tongue? I think not.

Actually, I know not, because as soon as I saw that ridiculous quote I ordered some more McKeith books. Not only does she make the same mistake in numerous other places, but it seems to me that her understanding of even the most basic elements of science is deeply, strangely distorted. In *You Are What You Eat* (p.211) she says: 'Each sprouting seed is packed with the nutritional energy needed to create a full grown healthy plant.'

This is hard to follow. Does a fully grown, healthy oak tree, a hundred feet tall, contain the same amount of energy as a tiny acorn? No. Does a fully grown, healthy sugarcane plant contain the same amount of nutritional energy – measure it in 'calories' if you like – as a sugarcane seed? No. Stop me if I'm boring you, in fact stop me if I've misunderstood something in what she's said, but to me this seems like almost the same mistake as the photosynthesis thing, because that extra energy to grow a fully grown plant comes, again, from photosynthesis, where plants use light to turn carbon dioxide and water into sugar and then into everything else that plants are made of.

This is not an incidental issue, an obscure backwater of McKeith's work, nor is it a question of which 'school of thought' you speak for: the 'nutritional energy' of a piece of food is one of the most important things you could possibly think of for a nutritionist to know about. I can tell you for a fact that the amount of nutritional energy you will get from eating one sugarcane seed is a hell of a lot less than you'd get from eating all the sugarcane from the plant that grew from it. These aren't passing errors, or slips of the tongue (I have a policy, as it were, of not quibbling on spontaneous utterances, because we all deserve the chance to fluff): these are clear statements from published tomes.

Watching McKeith's TV show with the eye of a doctor, it rapidly becomes clear that even here, frighteningly, she doesn't seem to know what she's talking about. She examines patients' abdomens on an examination couch as if she is a doctor, and confidently announces that she can feel which organs are inflamed. But clinical examination is a fine art at the best of times, and what she is claiming is like identifying which fluffy toy someone has hidden under a mattress (you're welcome to try this at home).

She claims to be able to identify lymphoedema, swollen ankles from fluid retention, and she almost does it right – at least, she kind of puts her fingers in roughly the right place, but only for about half a second, before triumphantly announcing her findings. If you'd like to borrow my second edition copy of Epstein and de Bono's *Clinical Examination* (I don't think there were many people in my year at medical school who didn't buy a copy), you'll discover that to examine for lymphoedema, you press firmly for around thirty seconds, to gently compress the exuded fluid out of the tissues, then take your fingers away, and look to see if they have left a dent behind.

In case you think I'm being selective, and only quoting McKeith's most ridiculous moments, there's more: the tongue is 'a window to the organs – the right side shows what the gall-bladder is up to, and the left side the liver'. Raised capillaries on your face are a sign of 'digestive enzyme insufficiency – your body is screaming for food enzymes'. Thankfully, Gillian can sell you some food enzymes from her website. 'Skid mark stools' (she is obsessed with faeces and colonic irrigation) are 'a sign of dampness inside the body – a very common condition in Britain'. If your stools are foul-smelling you are 'sorely in need of digestive enzymes'. Again. Her treatment for pimples on the forehead – not pimples anywhere else, mind you, only on the forehead – is a regular enema. Cloudy urine is 'a sign that your body is damp and acidic, due to eating the wrong foods'. The spleen is 'your energy battery'.

So we have seen scientific facts – very basic ones – on which Dr McKeith seems to be mistaken. What of scientific process? She has claimed, repeatedly and to anyone who will listen, that she is engaged in clinical scientific research. Let's step back a moment, because from everything I've said, you might reason-ably assume that McKeith has been clearly branded as some kind of alternative therapy maverick. In fact, nothing could be further from the truth. This doctor has been presented, consistently, up front, by Channel 4, her own website, her management company and her books as a scientific authority on nutrition.

Many watching her TV show quite naturally assumed she was a medical doctor. And why not? There she was, examining patients, performing and interpreting blood tests, wearing a white coat, surrounded by test tubes, 'Dr McKeith', 'the diet doctor', giving diagnoses, talking authoritatively about treat-ment, using complex scientific terminology with all the author-ity she could muster, and sticking irrigation equipment nice and invasively right up into people's rectums.

Now, to be fair, I should mention something about the doctorate, but I should also be clear: I don't think this is the most important part of the story. It's the funniest and most memorable part of the story, but the real action is whether McKeith is capable of truly behaving like the nutritional science academic she claims to be.

And the scholarliness of her work is a thing to behold. She produces lengthy documents that have an air of 'referenciness', with nice little superscript numbers, which talk about trials, and studies, and research, and papers ... but when you follow the numbers, and check the references, it's shocking how often they aren't what she claimed them to be in the main body of the text, or they refer to funny little magazines and books, such as *Delicious, Creative Living, Healthy Eating*, and my favourite, *Spiritual Nutrition and the Rainbow Diet*, rather than proper academic journals.

She even does this in the book *Miracle Superfood*, which, we are told, is the published form of her PhD. 'In laboratory experiments with anaemic animals, red-blood cell counts have returned to normal within four or five days when chlorophyll was given,' she says. Her reference for this experimental data is a magazine called *Health Store News*. 'In the heart,' she explains, 'chlorophyll aids in the transmission of nerve impulses that control contraction.' A statement that is referenced to the second issue of a magazine called *Earthletter*. Fair enough, if that's what you want to read – I'm bending over to be reasonable here – but it's clearly not a suitable source to reference that claim. This is her PhD, remember.

To me this is cargo-cult science, as Professor Richard Feynman described it over thirty years ago, in reference to the similarities between pseudoscientists and the religious activities on a few small Melanesian islands in the 1950s:

> During the war they saw aeroplanes with lots of good materials, and they want the same thing to happen now. So they've arranged to make things like runways, to put fires along the sides of the runways, to make a wooden hut for a man to sit in, with two wooden pieces on his head as headphones and bars of bamboo sticking out like antennas – he's the controller – and they wait for the aeroplanes to land. They're doing everything right. The form is perfect. It looks exactly the way it looked before. But it doesn't work. No aeroplanes land.

Like the rituals of the cargo cult, the form of McKeith's pseudo-academic work is superficially correct: the superscript numbers are there, the technical words are scattered about, she talks about research and trials and findings – but the substance is lacking. I actually don't find this very funny. It makes me quite depressed to think about her, sitting up, perhaps alone, studiously and earnestly typing this stuff out.

Should you feel sorry for her? One window into her world is the way in which she has responded to criticism: with statements that seem to be, well, wrong. It's cautious to assume that she will do the same thing with anything that I write here, so in preparation for the rebuttals to come, let's look at some of the rebuttals from the recent past.

In 2007, as has been noted, she was censured by the MHRA for selling a rather crass range of herbal sex pills called Fast Formula Horny Goat Weed Complex, advertised as having been shown by a 'controlled study' to promote sexual satisfaction, and sold with explicit medicinal claims. They were illegal for sale in the UK. She was ordered to remove the products from sale immediately. She complied – the alternative would have been prosecution – but her website announced that the sex pills had been withdrawn because of 'the new EU licensing laws regarding herbal products'. She engaged in a spot of Europhobic banter with the *Scottish Herald* newspaper: 'EU bureaucrats are clearly concerned that people in the UK are having too much good sex,' she explained.

Nonsense. I contacted the MHRA, and they said: 'This has nothing to do with new EU regulations. The information on the McKeith website is incorrect.' Was it a mistake? 'Ms McKeith's organisation had already been made aware of the requirements of medicines legislation in previous years; there was no reason at all for all the products not to be compliant with the law.' They went on. 'The Wild Pink Yam and Horny Goat Weed products marketed by McKeith Research Ltd were never legal for sale in the UK.'

Then there is the matter of the CV. Dr McKeith's PhD is from Clayton College of Natural Health, a non-accredited correspondence course college, which unusually for an academic institution also sells its own range of vitamin pills through its website. Her masters degree is from the same august institution. At current Clayton prices, it's $6,400 in fees for the PhD, and less for the masters, but if you pay for both at once you get a $300 discount (and if you really want to push the boat out, they have a package deal: two doctorates and a masters for $12,100 all in).

On her CV, posted on her management website, McKeith claimed to have a PhD from the rather good American College of Nutrition. When this was pointed out, her representative explained that this was merely a mistake, made by a Spanish work experience kid who posted the wrong CV. The attentive reader may have noticed that the very same claim about the American College of Nutrition was also in one of her books from several years previously.

In 2007 a regular from my website – I could barely contain my pride – took McKeith to the Advertising Standards Authority, complaining about her using the title 'doctor' on the basis of a qualification gained by correspondence course from a non-accredited American college: and won. The ASA came to the view that McKeith's advertising breached two clauses of the Committee of Advertising Practice code: 'substantiation' and 'truthfulness'.

Dr McKeith sidestepped the publication of a damning ASA draft adjudication at the last minute by accepting – 'voluntarily' – not to call herself 'doctor' in her advertising any more. In the news coverage that followed, McKeith suggested that the adjudication was only concerned with whether she had presented herself as a medical doctor. Again, this is not true. A copy of that draft adjudication has fallen into my lap – imagine that – and it specifically says that people seeing the adverts would reasonably expect her to have either a medical degree, or a PhD from an accredited university.

She even managed to get one of her corrections into a profile on her in my own newspaper, the *Guardian*: 'Doubt has also been cast on the value of McKeith's certified membership of the American Association of Nutritional Consultants, especially since *Guardian* journalist Ben Goldacre managed to buy the same membership online for his dead cat for $60. McKeith's spokeswoman says of this membership: "Gillian has 'professional membership', which is membership designed for practising nutritional and dietary professionals, and is distinct from 'associate membership', which is open to all individuals. To gain professional membership Gillian provided proof of her degree and three professional references." '

Well. My dead cat Hettie is also a 'certified professional member' of the AANC. I have the certificate hanging in my loo. Perhaps it didn't even occur to the journalist that McKeith could be wrong. More likely, in the tradition of nervous journalists, I suspect that she was hurried, on deadline, and felt she had to get McKeith's 'right of reply' in, even if it cast doubts on – I'll admit my beef here – my own hard-won investigative revelations about my dead cat. I mean, I don't sign my dead cat up to bogus professional organisations for the good of my health, you know. It may sound disproportionate to suggest that I will continue to point out these obfuscations for as long as they are made, but I will, because to me, there is a strange fascination in tracking their true extent.

Although perhaps I should not be so bold. She has a libel case against the *Sun* over comments it made in 2004. The *Sun* is part of a large, wealthy media conglomerate, and it can protect itself with a large and well-remunerated legal team. Others can't. A charming but obscure blogger called PhDiva made some relatively innocent comments about nutritionists, mentioning McKeith, and received a letter threatening costly legal action from Atkins Solicitors, 'the reputation and brand-management specialists'. Google received a threatening legal letter simply for linking to – forgive me – a fairly obscure webpage on McKeith. She has also made legal threats to an excellently funny website called Eclectech for featuring an animated cartoon of her singing a silly song at around the time she was on *Fame Academy*.

Most of these legal tussles revolve around the issue of her qualifications, but such things shouldn't be difficult or complicated. If anyone wanted to check my degrees, memberships or affiliations, they could call up the institutions concerned and get instant confirmation: job done. If you said I wasn't a doctor, I wouldn't sue you; I'd roar with laughter.

But if you contact the Australasian College of Health Sciences (Portland, Oregon), where McKeith has a 'pending diploma in herbal medicine', they say they can't tell you anything about their students. If you contact Clayton College of Natural Health to ask where you can read her PhD, they say you can't. What kind of organisations are these? If I said I had a PhD from Cambridge, US or UK (I have neither, and I claim no great authority), it would only take you a day to find it in their library.

These are perhaps petty episodes. But for me the most concerning aspect of the way she responds to questioning of her scientific ideas is exemplified by a story from 2000, when Dr McKeith approached a retired Professor of Nutritional Medicine from the University of London. Shortly after the publication of her book *Living Food for Health*, John Garrow wrote an article about some of the bizarre scientific claims Dr McKeith was making, and his piece was published in a fairly obscure medical newsletter. He was struck by the strength with which she presented her credentials as a scientist ('I continue every day to research, test and write furiously so that you may benefit …' etc.). He has since said that he assumed – like many others – that she was a proper doctor. Sorry: a medical doctor. Sorry: a qualified, conventional medical doctor who attended an accredited medical school.

In this book McKeith promised to explain how you can 'boost your energy, heal your organs and cells, detoxify your body, strengthen your kidneys, improve your digestion, strengthen your immune system, reduce cholesterol and high blood pressure, break down fat, cellulose and starch, activate the enzyme energies of your body, strengthen your spleen and liver function, increase mental and physical endurance, regulate your blood sugar, and lessen hunger cravings and lose weight'.

These are not modest goals, but her thesis was that they were all possible with a diet rich in enzymes from 'live' raw food – fruit, vegetables, seeds, nuts, and especially live sprouts, which are 'the food sources of digestive enzymes'. She even offered 'combination living food powder for clinical purposes', in case people didn't want to change their diet, and explained that she used this for 'clinical trials' with patients at her clinic.

Garrow was sceptical of her claims. Apart from anything else, as Emeritus Professor of Human Nutrition at the University of London, he knew that human animals have their own digestive enzymes, and any plant enzyme you eat is likely to be digested like any other protein. As any Professor of Nutrition, and indeed many GCSE biology students, could tell you.

Garrow read McKeith's book closely, as have I. These 'clinical trials' seemed to be a few anecdotes about how incredibly well her patients felt after seeing her. No controls, no placebo, no attempt to quantify or measure improvements. So Garrow made a modest proposal in a fairly obscure medical newsletter. I am quoting it in its entirety, partly because it is a rather elegantly written exposition of the scientific method by an eminent academic authority on the science of nutrition, but mainly because I want you to see how politely he stated his case:

> I also am a clinical nutritionist, and I believe that many of the statements in this book are wrong. My hypothesis is that any benefits which Dr McKeith has observed in her patients who

take her living food powder have nothing to do with their enzyme content. If I am correct, then patients given powder which has been heated above 118°F for twenty minutes will do just as well as patients given the active powder. This amount of heat would destroy all enzymes, but make little change to other nutrients apart from vitamin C, so both groups of patients should receive a small supplement of vitamin C (say 60mg/day). However, if Dr McKeith is correct, it should be easy to deduce from the boosting of energy, etc., which patients received the active powder and which the inactivated one.

Here, then, is a testable hypothesis by which nutritional science might be advanced. I hope that Dr McKeith's instincts, as a fellow-scientist, will impel her to accept this challenge. As a further inducement I suggest we each post, say, £1,000, with an independent stakeholder. If we carry out the test, and I am proved wrong, she will of course collect my stake, and I will publish a fulsome apology in this newsletter. If the results show that she is wrong I will donate her stake to HealthWatch [a medical campaigning group], and suggest that she should tell the 1,500 patients on her waiting list that further research has shown that the claimed benefits of her diet have not been observed under controlled conditions. We scientists have a noble tradition of formally withdrawing our publications if subsequent research shows the results are not reproducible – don't we?

Sadly, McKeith – who, to the best of my knowledge, despite all her claims about her extensive 'research', has never published in a proper 'Pubmed-listed' peer-reviewed academic journal – did not take up this offer to collaborate on a piece of research with a Professor of Nutrition. Instead, Garrow received a call from McKeith's lawyer husband, Howard Magaziner, accusing him of defamation and promising legal action. Garrow, an immensely affable and relaxed old academic, shrugged this off with style.

He told me, 'I said, "Sue me." I'm still waiting.' His offer of £1,000 still stands.

But there is one vital issue we have not yet covered. Because despite the way she seems to respond to criticism or questioning of her ideas, her illegal penis pills, the unusually complicated story of her qualifications, despite her theatrical abusiveness, and the public humiliation pantomime of her shows, in which the emotionally vulnerable and obese cry on television, despite her apparently misunderstanding some of the most basic aspects of GCSE biology, despite doling out 'scientific' advice in a white coat, despite the dubious quality of the work she presents as somehow being of 'academic' standard, despite the unpleasantness of the food she endorses, there are still many who will claim: 'You can say what you like about McKeith, but she has improved the nation's diet.'

This is not to be shrugged off lightly. Let me be very clear, for I will say it once again: anyone who tells you to eat more fresh fruit and vegetables is all right by me. If that was the end of it, I'd be McKeith's biggest fan, because I'm all in favour of 'evidence-based interventions to improve the nation's health', as they used to say to us in medical school.

Let's look at the evidence. Diet has been studied very extensively, and there are some things that we know with a fair degree of certainty: there is reasonably convincing evidence that having a diet rich in fresh fruit and vegetables, with natural sources of dietary fibre, avoiding obesity, moderating one's intake of alcohol, cutting out cigarettes and taking physical exercise are protective against things such as cancer and heart disease.

Nutritionists don't stop there, because they can't: they have to manufacture complication, to justify the existence of their profession. These new nutritionists have a major commercial problem with the evidence. There's nothing very professional or proprietary about 'Eat your greens,' so they have had to push things further. But unfortunately for them, the technical, confusing, overcomplicated, tinkering interventions that they promote – the enzymes, the exotic berries – are very frequently not supported by convincing evidence.

That's not for lack of looking. This is not a case of the medical hegemony neglecting to address the holistic needs of the people. In many cases the research has been done, and has shown that the more specific claims of nutritionists are actually wrong. The fairy tale of antioxidants is a perfect example. Sensible dietary practices, which we all know about, still stand. But the unjustified, unnecessary overcomplication of this basic dietary advice is, to my mind, one of the greatest crimes of the nutritionist movement. As I have said, I don't think it's excessive to talk about consumers paralysed with confusion in supermarkets.

But it's just as likely that they will be paralysed with fear. They may have a bad reputation for paternalism, but it's hard to picture any doctor in the past century using McKeith's consultation methods as a serious tactic for inducing lifestyle change in his patients. With McKeith we see fire and brimstone hailing down until her subjects cry on national television: a chocolate gravestone with your name on it in the garden; a shouty dressing-down in public for the obese. As a posture it is as seductive as it is telegenic, it has a sense of generating movement; but if you drag yourself away from the theatricality of souped-up recipe and lifestyle shows on telly, the evidence suggests that scare campaigns may not get people changing their behaviour in the long term.

What can you do? There's the rub. The most important take-home message with diet and health is that anyone who ever expresses anything with certainty is basically wrong, because the evidence for cause and effect in this area is almost always weak and circumstantial, and changing an individual person's diet may not even be where the action is.

What is the best evidence on the benefits of changing an individual person's diet? There have been randomised controlled trials, for example – where you take a large group of people, change their diet, and compare their health outcomes with another group – but these have generally shown very disappointing results.

The Multiple Risk Factor Intervention Trial was one of the largest medical research projects ever undertaken in the history of mankind, involving over 12,866 men at risk of cardiovascular events, who went through the trial over seven years. These people were subjected to a phenomenal palaver: questionnaires, twenty-four-hour dietary recall interviews, three-day food records, regular visits, and more. On top of this, there were hugely energetic interventions which were supposed to change the lives of individuals, but which by necessity required that whole families' eating patterns were transformed: so there were weekly group information sessions for participants – and their wives – individual work, counselling, an intensive education programme, and more. The results, to everyone's disappointment, showed no benefit over the control group (who were not told to change their diet). The Women's Health Initiative was another huge randomised controlled trial into dietary change, and it gave similarly gave negative results. They all tend to.

Why should this be? The reasons are fascinating, and a window into the complexities of changing health behaviour. I can only discuss a few here, but if you are genuinely interested in preventive medicine – and you can cope with uncertainty and the absence of quick-fix gimmicks – then may I recommend you pursue a career in it, because you won't get on telly, but you will be both dealing in sense and doing good.

The most important thing to notice is that these trials require people to turn their entire lives upside down, and for about a decade. That's a big ask: it's hard enough to get people signed up for participating in a seven-week trial, let alone one that lasts seven years, and this has two interesting effects. Firstly, your participants probably won't change their diets as much as you want them to; but far from being a failing, this is actually an excellent illustration of what happens in the real world: individual people do not, in reality, change their diets at the drop of a hat, alone, as individuals, for the long term. A dietary change probably requires a change in lifestyle, shopping habits, maybe even what's in the shops, how you use your time, it might even require that you buy some cooking equipment, how your family relates to each other, change your work style, and so on.

Secondly, the people in your 'control group' will change their diets too: remember, they've agreed voluntarily to take part in a hugely intrusive seven-year-long project that could require massive lifestyle changes, so they may have a greater interest in health than the rest of your population. More than that, they're also being weighed, measured, and quizzed about their diet, all at regular intervals. Diet and health are suddenly much more at the forefront of their minds. They will change too.

This is not to rubbish the role of diet in health – I bend over backwards to find some good in these studies – but it does reflect one of the most important issues, which is that you might not start with goji berries, or vitamin pills, or magic enzyme powders, and in fact you might not even start with an individual changing their diet. Piecemeal individual life changes – which go against the grain of your own life and your environment – are hard to make, and even harder to maintain. It's important to see the individual – and the dramatic claims of all lifestyle nutritionists, for that matter – in a wider social context.

Reasonable benefits have been shown in intervention studies – like the North Karelia Project in Finland – where the public health gang have moved themselves in lock, stock and barrel to set about changing everything about an entire community's behaviour, liaising with businesses to change the food in shops, modifying whole lifestyles, employing community educators and advocates, improving healthcare provision, and more, producing some benefits, if you accept that the methodology used justifies a causal inference. (It's tricky to engineer a control group for this kind of study, so you have to make pragmatic decisions about study design, but read it online and decide for yourself: I'd call it a 'large and promising case study'.)

There are fairly good grounds to believe that many of these lifestyle issues are, in fact, better addressed at the societal level. One of the most significant 'lifestyle' causes of death and disease, after all, is social class. To take a concrete example, I rent a flat in London's Kentish Town on my modest junior medic's salary (don't believe what you read in the papers about doctors' wages). This is largely a white, working-class area, and the adult male life expectancy is about seventy years. Two miles away, in Hampstead, where the millionaire entrepreneur Dr Gillian McKeith PhD owns a large property, surrounded by other wealthy middle-class people, male life expectancy is almost eighty years. I know this because I have the Annual Public Health Report for Camden open on my kitchen table right now.

The reason for this phenomenal disparity in life expectancy – the difference between a lengthy and rich retirement, and a very truncated one indeed – is not that the people in Hampstead are careful to eat goji berries and a handful of Brazil nuts every day, thus ensuring they're not deficient in selenium, as per nutritionists' advice. That's a fantasy, and in some respects one of the most destructive features of the whole nutritionist project, graphically exemplified by McKeith: it's a distraction from the real causes of ill health, but also – do stop me if I'm pushing this too far – in some respects, a manifesto of right-wing individualism. You are what you eat, and people die young because they deserve it. *They* choose death, through ignorance and laziness, but *you* choose life, fresh fish, olive oil, and that's why you're healthy. You're going to see eighty. You deserve it. Not like *them*.

Back in the real world, genuine public health interventions to address the social and lifestyle causes of disease are far less lucrative, and far less of a spectacle, than anything a Gillian McKeith – or, more importantly, a television commissioning editor – would ever dream of dipping into. What prime-time TV series looks at food deserts created by giant supermarket chains, the very companies with which these stellar media nutritionists so often have their lucrative commercial contracts? Who puts the issue of social inequality driving health inequality onto our screens? Where's the human interest in prohibiting the promotion of bad foods, facilitating access to healthier foods by means of taxation, or maintaining a clear labelling system?

Where is the spectacle in 'enabling environments' that naturally promote exercise, or urban planning that prioritises cyclists, pedestrians and public transport over the car? Or in reducing the ever-increasing inequality between senior executive and shop-floor pay? When did you ever hear about elegant ideas like 'walking school buses', or were stories about their benefits crowded out by the latest urgent front-page food fad news?

I don't expect Dr Gillian McKeith, or anyone in the media, to address a single one of these issues, and neither do you: because if we are honest with ourselves we understand that these programmes are only partly about food, and much more about salacious and prurient voyeurism, tears, viewing figures and vaudeville.

Dr McKeith puts a cabbie straight

Here is my favourite Dr McKeith story, and it comes from her own book, *Living Food for Health*. She is in a cab, and the driver, Harry, has spotted her. He tries to spark up a friendly conversation by suggesting that fish contains more omega oils than flax. Dr McKeith disputes this: 'Flax seeds contain far greater levels of the healthy oils (omega-3 and omega-6) in a properly balanced and assimilable form.' When Harry disagrees, she replies: 'What do you mean, you disagree? Have you spent years conducting clinical research, working with patients, lecturing, teaching, studying the omega oils in flax, obtaining worldwide data, compiling one of the largest private health libraries on the planet, and writing extensively on the topic? Are you a scientist, a biochemist, a botanist, or have you spent a lifetime studying food and biochemistry as I have done? Where is your scientific authority?' Harry responds that his wife is a doctor, a gynaecologist. 'Is she a food specialist or nutritional biochemist as well?' demands Dr McKeith. 'Um, ah, well, no, but she is a doctor.'

I am not a food specialist, nor am I a nutritional biochemist. In fact, as you know, I claim no special expertise whatsoever: I hope I can read and critically appraise medical academic literature – something common to all recent medical graduates – and I apply this pedestrian skill to the millionaire businesspeople who drive our culture's understanding of science.

Flax seeds contain large amounts of fibre (along with oestrogenic compounds), so they're not very 'assimilable', as Dr McKeith asserts, unless you crush them, in which case they taste foul. They're sold as a laxative in doses of 15g, and you will need a lot of the stuff, partly because there's also a problem with the form of omega oils in them: flax contains a short-chain plant form, and these must be converted in your body to the long-chain animal forms which may be beneficial (called DHA and EPA). When you account for the poor conversion in the body, then flax seeds and fish contain roughly the same amounts of omega oil.

We must also remember that we live not in a laboratory, but in the real world. It's very easy to eat 100g of mackerel – if this were a completely different kind of book I'd be giving you my kedgeree recipe right now – whereas I would suggest that it's slightly trickier to get a tablespoon of flax seed into you. Parsley, similarly, is a rich source of vitamin C, but you're not going to eat an orange-sized lump of the stuff. As for Dr McKeith's further claim that flax is 'properly balanced', I don't know if she means spiritually or biologically, but fish is much higher in omega-3, which most people would say is better.

More importantly, why is everyone talking about omega-3? On to the next chapter.

8

'Pill Solves Complex Social Problem'

Medicalisation – or 'Will fish-oil pills make my child a genius?'

In 2007 the *British Medical Journal* published a large, well-conducted, randomised controlled trial, performed at lots of different locations, run by publicly funded scientists, and with a strikingly positive result: it showed that one treatment could significantly improve children's antisocial behaviour. The treatment was entirely safe, and the study was even accompanied by a very compelling cost-effectiveness analysis.

Did this story get reported as front-page news in the *Daily Mail*, the natural home of miracle cures (and sinister hidden scares)? Was it followed up on the health pages, with an accompanying photo feature, describing one child's miraculous recovery, and an interview with an attractive relieved mum with whom we could all identify?

No. The story was unanimously ignored by the British news media, despite their preoccupation with both antisocial behaviour and miracle cures, for one simple reason: the research was not about a pill. It was about a cheap, practical parenting programme.

At the same time, for over five years now, newspapers and television stations have tried to persuade us, with 'science', that fish-oil pills have been proven to improve children's school performance, IQ, behaviour, attention, and more. In fact, nothing could be further from the truth. We are about to learn some very interesting lessons about the media, about how not to conduct a trial, and about our collective desire for medicalised, sciencey-sounding explanations for everyday problems. Do fish-oil pills work? Do they make your child cleverer and better-behaved? The simple answer is, at the moment, nobody could possibly know. Despite everything you have been told, a trial has never been done in mainstream children.

The newspapers would have you believe otherwise. I first became aware of 'the Durham trials' when I saw on the news that a trial of fish-oil capsules was being planned in 5,000 children. It's an astonishing testament to the news values of the British media that this piece of research remains, I am quite prepared to suggest, probably the single most well-reported clinical trial of the past few years. It was on Channel 4 and ITV, and in every national newspaper, sometimes repeatedly. Impressive results were confidently predicted.

This rang alarm bells for two reasons. Firstly, I knew the results of the previous trials of fish-oil capsules in children – I'll describe them in due course – and they weren't very exciting. But more than that, as a basic rule, I would say this: whenever somebody tells you that their trial is going to be positive before they've even started it, then you know you're onto an interesting story.

Here is what they were planning to do in their 'trial': recruit 5,000 children in their GCSE year, give them all six fish-oil capsules a day, then compare how they did in their exams with how the council had estimated they should do without the capsules. There was no 'control' group to compare against (like the Aqua Detox bath with no feet in it, or the ear candle on the picnic table, or a group of children taking placebo capsules without fish oil in them). Nothing.

By now you might not need me to tell you that this is a preposterous and above all wasteful way to go about doing a study on a pill that is supposed to improve school performance, with £1 million worth of generously donated capsules and 5,000 children at your disposal. But humour an old man, and let me flesh out your hunches, because if we cover the theoretical issues properly first, then the 'researchers' at Durham become even more amusingly absurd.

Why you have a placebo group

If you divide a group of kids in half, and give a placebo capsule to one group, and the real capsule to the other group, you can then compare how well each group does, and see whether it was the ingredients in the pill that made the difference to their performance, or just the fact of taking a pill, and being in a study. Why is this important? Because you have to remember that whatever you do to children, in a trial of a pill to improve their performance, their performance will improve.

Firstly, children's skills improve over time anyway: they grow up, time passes, and they get better at stuff. You might think you're clever, sitting there with no nappy on, reading this book, but things haven't always been that way, as your mother could tell you.

Secondly, the children – and their parents – know that they are being given these tablets to improve their performance, so they will be subject to a placebo effect. I have already harped on about this at phenomenal length, because I think the real scientific story of the connections between body and mind are infinitely more interesting than anything concocted by the miracle-cure community, but here it is enough to remind you that the placebo effect is very powerful: consciously or unconsciously, the children will expect themselves to improve, and so will their parents and their teachers. Children are exquisitely sensitive to our expectations of them, and anyone who doubts that fact should have their parenting permit revoked.

Thirdly, children will do better just from being in a special group that is being studied, observed and closely attended to, since it seems that the simple fact of *being in a trial* improves your performance, or recovery from illness. This phenomenon is called the 'Hawthorne effect', not after a person, but after the factory where it was first observed. In 1923 Thomas Edison (he of the lightbulb) was chairing the 'Committee on the Relation of Quality and Quantity of Illumination to Efficiency in the Industries'. Various reports from several companies had suggested that better lighting might increase productivity, so a researcher called Deming went with his team to test the theory at Western Electric's Hawthorne plant at Cicero, Illinois.

I will give you the simplified 'myth' version of the findings, as a rare compromise between pedantry and simplicity. When the researchers increased light levels, they found that performance improved. But when they reduced the light levels, performance improved then, too. In fact, they found that no matter what they did, productivity increased anyway. This finding was very important: when you tell workers they are part of a special study to see what might improve productivity, and then you do something … they improve their productivity. This is a kind of placebo effect, because the placebo is not about the mechanics of a sugar pill, it is about the cultural meaning of an intervention, which includes, amongst other things, your expectations, and the expectations of the people tending to you and measuring you.

Beyond all this technical stuff, we also have to place the GCSE results – the outcome which was being measured in this 'trial' – in their proper context. Durham has a very bad exam record, so its council will be battling hard in every way it possibly can to improve school performance, with all manner of different initiatives and special efforts, and extra funding, running simultaneously with the fish-oil 'trials'.

We should also remember that bizarre English ritual whereby GCSE results get better every year, yet anyone who suggests that the exams are getting easier is criticised for undermining the achievement of the successful candidates. In fact, taking the long view, this easing is obvious: there are forty-year-old O-level papers which are harder than the current A-level syllabus; and there are present-day university finals papers in maths that are easier than old A-level papers.

To recap: GCSE results will get better anyway; Durham will be desperately trying to improve its GCSE results through other methods anyway; and any kids taking pills will improve their GCSE results anyway, because of the placebo effect and the Hawthorne effect.

This could all be avoided by splitting the group in half and giving a placebo to one group, separating out what is a specific effect of the fish-oil pills, and what is a general effect of all the other stuff we've described above. That would give you very useful information.

Is it ever acceptable to do the kind of trial that was being carried out in Durham? Yes. You can do something called an 'open trial', without a placebo group, and this is an accepted kind of research. In fact, there is an important lesson about science here: you can do a less rigorous experiment, for practical reasons, as long as you are clear about what you are doing when you present your study, so other people can make their own minds up about how they interpret your findings.

But there is an important caveat. If you do this kind of 'compromise' study, without a placebo group, 'open label', but in the hope of getting the most accurate possible picture of the treatment's benefits, then you do it as carefully as you can, while being fully aware that your results might be distorted by expectation, by the placebo effect, by the Hawthorne effect, and so on. You might sign up your kids calmly and cautiously, saying in a casual, offhand fashion that you're doing a small informal study on some tablets, you don't say what you expect to find, you hand them out without fanfare, and you calmly measure the results at the end.

What they did in Durham was the polar opposite. There were camera crews, soundmen and lighting men flooding the classrooms. Children were interviewed for radio, for television, for the newspapers; so were their parents; so were their teachers: so were Madeleine Portwood, the educational psychologist who was performing the trial, and Dave Ford, Head of Education, talking – bizarrely – about how they confidently expected positive results. They did literally everything which, in my view, would guarantee them a false positive result, and ruin any chance of their study giving meaningful and useful new information. How often does this happen? In the world of nutritionism, sadly, it seems to be standard research protocol.

We should also remember that these fish-oil 'trials' were measuring some highly volatile outcomes. Performance at school in a test, and 'behaviour' (a word with a large semantic footprint, if ever I saw one) are huge, variable, amorphous things. More than most outcomes, they will change from moment to moment, with different circumstances, frames of mind, and expectations. Behaviour is not like a blood haemoglobin level, or even height, and nor is intelligence.

Durham Council and Equazen were so successful in their publicity drive, whether through an uncontainable enthusiasm for a positive result or simple foolishness (I really don't know which) that they effectively sabotaged their 'trial'. Before the first fish-oil capsule was swallowed by a single child, the Eye Q branded supplement and trial had received glowing publicity in the local papers, the *Guardian*, the *Observer*, the *Daily Mail*, *The*

Times, Channel 4, the BBC, ITV, the *Daily Express*, the *Daily Mirror*, the *Sun*, GMTV, *Woman's Own*, and many more. Nobody could claim that the children weren't well primed.*

You're not an educational psychologist. You're not the Head of Education at a council. You're not the long-standing MD of a multi-million-pound pill business running huge numbers of 'trials'. But I am quite sure that you understand very clearly all of these criticisms and concerns, because this isn't rocket science.

Durham defend themselves

Being a fairly innocent and open-minded soul, I went to the people behind the trial, and put it to them that they had done the very things which would guarantee that their trial would produce useless results. That is what anyone would do in an academic context, and this was a trial after all. Their response was simple. 'We've been quite clear,' said Dave Ford, Chief Schools Inspector for Durham, and the mastermind behind the project to give out the capsules and measure the results. 'This is not a trial.'

This felt a bit weak. I call up to suggest that they're doing a badly designed piece of research, and suddenly everything's OK, because it's not actually a 'trial'? There were other reasons for thinking this was a fairly implausible defence. The Press Association called it a trial. The *Daily Mail* called it a trial. Channel 4 and ITV and everyone covering it all presented it, very clearly, as research (you can see the clips at badscience.net). More importantly, Durham Council's own press release called it

---

* In fact, it's hard to overstate quite how large the fish-oil circus became, over the many years it ran. Professor Sir Robert Winston himself, moustachioed presenter of innumerable 'science' programmes for the BBC, personally endorsed a competing omega-3 product, in an advertising campaign which was ultimately terminated by the ASA since it breached their codes on truthfulness and substantiation.

a 'study' and a 'trial', repeatedly.* They were giving something to schoolchildren and measuring the result. Their own descriptive term for this activity was 'trial'. Now they were saying it wasn't a trial.

I moved on to Equazen, the manufacturer which is still being lauded throughout the press for its involvement in these 'trials' which were almost guaranteed – by virtue of the methodological flaws we have already discussed – to produce spurious positive results. Adam Kelliher, chief executive of the company, clarified further: this was an 'initiative'. It was not a 'trial', nor was it a 'study', so I could not critique it as such. Although it was hard to ignore the fact that the Equazen press release talked about giving a capsule and measuring the results, and the word which the company itself used to describe this activity was: 'trial'.

Dr Madeleine Portwood, the senior educational psychologist running the study, called it a 'trial' (twice in the *Daily Mail*). Every single write-up described it as research. They were giving 'X' and measuring change 'Y'. They called it a trial, and it was a trial – but a stupid trial. Simply saying, 'Ah, but this is not a trial' didn't strike me as an adequate – nor indeed a particularly adult defence. They didn't seem to think a trial was even necessary, and Dave Ford explained that the evidence already shows fish oils are beneficial. Let's see.

The fish-oil evidence

Omega-3 oils are 'essential fatty acids'. They're called 'essential' because they're not made by the body (unlike glucose or vitamin D, for example), so you have to eat them. This is true of a lot of things, like many vitamins, for example, and it's one of

---

* As a testament to the astonishing foolishness of Durham Council, they've now even gone to the trouble of changing the wording of this press release on their website, as if that might address the gaping design flaws.

the many reasons why it's a good idea to eat a varied diet, pleasure being another.

They are found in fish oils, and – in slightly different form – in evening primrose oil, linseed oil and other sources. If you look at the flow charts in a biochemistry textbook you will see that there is a long list of functions which these molecules perform in the body: they are involved in constructing membranes, and also some of the molecules that are involved in communication between cells, for example during inflammation. For this reason some people think it might be useful to eat them in larger amounts.

I'm open to the idea myself, but there are good reasons to be sceptical, because there is a lot of history here. In the past, decades before the Durham 'trials', the field of essential fatty acid research has seen research fraud, secrecy, court cases, negative findings that have been hushed up, media misreporting on a massive scale, and some very striking examples of people using the media to present research findings direct to the public in order to circumvent regulators. We'll come back to that later.

There have been – count them – six trials to date on fish oil in children. Not one of these trials was done in 'normal' mainstream children: all of them have been done in special categories of children with one diagnosis or another – dyslexia, ADHD, and so on. Three of the trials had some positive findings, in some of the many things they measured (but remember, if you measure a hundred things in a study, a few of them will improve simply by chance, as we will see later), and three were negative. One, amusingly, found that the placebo group did better than the fish-oil group on some measures. They are all summarised online at badscience.net.

And yet: 'All of our research, both published and unpublished, shows that the Eye Q formula can really help enhance achievement in the classroom,' says Adam Kelliher, CEO of Equazen. All of it.

To take a statement like this seriously, we would have to read the research. I am not for one nanosecond accusing anybody of research fraud: in any case, if anyone did suspect fraud, reading the research would not help, because if people have faked their results with any enthusiasm then you need a forensic statistician and a lot of time and information to catch them out. But we do need to read published research in order to establish whether the conclusions drawn by the stakeholders in the research are valid, or whether there are methodological problems that make their interpretation the product of wishful thinking, incompetence, or perhaps even simply a judgement call with which you would not concur.

Paul Broca, for example, was a famous French craniologist in the nineteenth century whose name is given to Broca's area, the part of the frontal lobe involved in the generation of speech (which is wiped out in many stroke victims). Among his other interests, Broca used to measure brains, and he was always rather perturbed by the fact that the German brains came out a hundred grams heavier than French brains. So he decided that other factors, such as overall body weight, should also be taken into account when measuring brain size: this explained the larger Germanic brains to his satisfaction. But for his prominent work on how men have larger brains than women, he didn't make any such adjustments. Whether by accident or by design, it's a kludge.

Cesare Lombroso, a nineteenth-century pioneer of 'biological criminology', made similarly inconsistent fixes in his research, citing insensitivity to pain among criminals and 'lower races' as a sign of their primitive nature, but identifying the very same quality as evidence of courage and bravery in Europeans. The devil is in the detail, and this is why scientists report their full methods and results in academic papers, not in newspapers or on television programmes, and it is why experimental research cannot be reported in the mainstream media alone.

You might feel, after the 'trial' nonsense, that we should be cautious about accepting Durham and Equazen's appraisal of their own work, but I'd be suspicious of the claims from a lot of serious academics in exactly the same way (they would welcome such suspicion, and I'd be able to read the research evidence they were drawing on). I asked Equazen for their twenty positive studies, and was told I would have to sign a confidentiality agreement to see them. That's a confidentiality agreement to review the research evidence for widely promoted claims, going back years, in the media and by Durham Council employees, about a very controversial area of nutrition and behaviour, of huge interest to the public, and about experiments conducted – forgive me if I'm getting sentimental here – on our schoolchildren. I refused.*

Meanwhile, all over the newspapers and television, everywhere you looked, going back to at least 2002, there were reports of positive fish-oil trials in Durham, using Equazen products. There seemed to have been half a dozen of these trials, in various locations, performed by Durham Council staff on Durham state school children, yet there was no sign of anything in the scientific literature (beyond one study by a researcher from Oxford, which was on children with developmental coordination disorder). There were wildly enthusiastic Durham Council press releases that talked about positive trials, sure. There were Madeleine Portwood interviews with the press, in which she talked enthusiastically about the positive results (and talked too about how the fish oil was improving children's skin conditions, and other problems) – but no published studies.

I contacted Durham. They put me on to Madeleine Portwood, the scientific brains behind this enormous and enduring

---

* Although you wouldn't know if I had signed, since I wouldn't be able to tell you.

operation. She appears regularly on telly talking about fish oils, using inappropriately technical terms like 'limbic' to a lay audience. 'It sounds complicated,' say TV presenters, 'but the science says …' Portwood is evidently very enthusiastic about talking to parents and journalists, but she did not return my calls. The press office took a week to reply to my emails. I asked for details of the studies they had performed, or were performing. The responses appeared to be inconsistent with the media coverage. At least one trial seemed to be missing. I asked for methodological details of the studies they were doing, and results of the ones that were completed. Not until we publish, they said.

Equazen and Durham Council had coached, preened and spoon-fed a huge number of journalists over the years, giving them time and energy; as far as I can see there is only one difference between me and those reporters: from what they wrote, they clearly know very little about trial design, whereas I know, well, a fair bit (and now so do you).

Meanwhile I kept being referred to the durhamtrials.org website, as if this contained useful data. It had evidently bamboozled many journalists and parents before me, and it's linked to by many news stories and Equazen adverts. But as a source of information about the 'trials', this site is a perfect illustration of why you should publish a trial before you make any dramatic claims about the results. It's hard to tell what's there. The last time I looked, there was some data borrowed from a proper trial published elsewhere by some Oxford researchers (which happened to be done in Durham), but other than that, no sign of Durham's own placebo-controlled trials which kept appearing in the news. There were plenty of complicated-looking graphs, but they seem to be on special Durham 'trials', with no placebo-control group. They seem to describe improvements, with sciencey-looking graphics to illustrate them, but there are no statistics to say if the changes were statistically significant.

It's almost impossible to express just how much data is missing from this site, and how useless that renders what is there. As an example, there is a 'trial' with its results reported in a graph, but nowhere on the entire site, as far as I can see, does it tell you how many children were in the study reported. It's hard to think of a single more basic piece of information than that. But you can find plenty of gushing testimonials that wouldn't be out of place on an alternative therapists' website selling miracle cures. One child says: 'Now I am not so interested in the TV. I just like reading books. The best place in all the world is the library. I absolutely love it.'

I felt the public deserved to know what had been done in these trials. This was probably the most widely reported clinical trial of the past few years, it was a matter of great public interest, and the experiments were performed on children by public servants. So I made a Freedom of Information Act request for the information you would need to know about a trial: what was done, who were the children, what measurements were taken, and so on. Everything, in fact, from the standardised and very complete 'CONSORT' guidelines which describe best practice in writing up trial results. Durham Council refused, on grounds of cost.

So I got readers of the column to ask for little bits of information, so that none of us was asking for anything very expensive. We were accused of running a 'vexatious' 'campaign' of 'harassment'. The head of the council complained about me to the *Guardian*. I was eventually told that my questions might be answered, if I travelled 275 miles north to Durham in person. Various readers appealed; they were told that much of the information they had been refused did not in any case exist.

Finally, in February 2008, after a disappointing fall in the rate of improvement in GCSE results, the council announced that there had never been any intention of measuring exam performance. This surprised even me. To be scrupulously precise, what they said, in answer to a written question from an indignant retired headmaster, was this: 'As we have said previously it was never intended, and the County Council never suggested, that it would use this initiative to draw conclusions about the effectiveness or otherwise of using Fish Oil to boost exam results.'

To say that this contradicts their earlier claims would be something of an understatement. In a *Daily Mail* article from 5 September 2006 headlined 'Fish Oil Study Launched to Improve GCSE Grades', Dave Ford, the council's Chief Schools Inspector, said: 'We will be able to track pupils' progress and measure whether their attainments are better than their predicted scores.' Dr Madeleine Portwood, senior educational psychologist running the 'trial', said: 'Previous trials have shown remarkable results and I am confident that we will see marked benefits in this one as well.'

Durham county council's own press release from the beginning of the 'trial' reads: 'Education chiefs in County Durham are to mount a unique back-to-school initiative today which they believe could result in record GCSE pass levels next summer.' It reports that children are being given pills 'to see whether the proven benefits it has already brought children and young people in earlier trials can boost exam performances too'. The council's Chief Schools Inspector is 'convinced' that these pills 'could have a direct impact on their GCSE results ... the county-wide trial will continue until the pupils complete their GCSE examinations next June, and the first test of the supplement's effectiveness will be when they sit their "mock" exams this December.' 'We are able to track pupils' progress and we can measure whether their attainments are better than their predicted scores,' says Dave Ford in the press release for the trial which, we are now told, was not a trial, and was never intended to collect any data on exam results. It was with some astonishment that I also noticed that they had changed their original press release on the Durham website, and removed the word 'trial'.

Why is all this important? Well, firstly, as I have said, this was the most well-reported trial of that year, and the fact that it was such a foolish exercise could only undermine the public's understanding of the very nature of evidence and research. When people realise that they are flawed by design, then exercises like this undermine the public's faith in research: this can only undermine willingness to participate in research, of course, and recruiting participants into trials is difficult enough at the best of times.

More than that, there are also some very important ethical issues here. People volunteer their bodies – and their children's bodies – to participate in trials, on the understanding that the results will be used to improve medical and scientific knowledge. They expect that research performed on them will be properly conducted, informative by design, and that the results will be published in full, for all to see.

I have seen the parent-information leaflets that were distributed for the Durham project, and they are entirely unambiguous in promoting the exercise as a scientific research project. The word 'study' is used seventeen times in one of these leaflets, although there is little chance that the 'study' (or 'trial', or 'initiative') can produce any useful data, for the reasons we have seen, and in any case it has now been announced that the effect on GCSE results will not be published.

For these reasons the trial was, in my opinion, unethical.* You will have your own view, but it is very hard to understand what justification there can be for withholding the results of

---

* While we're on the subject of ethics, Durham have claimed that to give placebo to half the children would *itself* be unethical: this is another very basic misunderstanding on their part. We do not know if fish oils are beneficial or not. That would be the point of doing some proper research into the area.

this 'trial' now that it has concluded. Educationalists, academic researchers, teachers, parents and the public should be permitted to review the methods and results, and draw their own conclusions on its significance, however weak the design was. In fact, this is the exact same situation as the data on antidepressants' efficacy being withheld by the drug companies, and a further illustration of the similarities between these pill industries, despite the food supplement pill industry's best efforts to present itself as somehow 'alternative'.

The power is in the pill?

We should be clear that I'm not – and I'm quite entitled to say this – myself very interested in whether fish-oil capsules improve children's IQ, and I say this for a number of reasons. Firstly, I'm not a consumer journalist, or a lifestyle guru, and despite the infinitely superior financial rewards, I am absolutely very much not in the business of 'giving readers health advice' (to be honest, I'd rather have spiders lay eggs in my skin). But also, if you think about it rationally, any benefit of fish oil for school performance will probably not be all that dramatic. We do not have an epidemic of thick vegetarians, for example, and humans have shown themselves to be as versatile as their diets are diverse, from Alaska to the Sinai desert.

But more than anything, at the risk of sounding like the most boring man you know, again: I wouldn't start with molecules, or pills, as a solution to these kinds of problems. I can't help noticing that the capsules Durham is promoting cost 80p per child per day, while it spends only 65p per child per day on school meals, so you might start there. Or you might restrict junk-food advertising to children, as the government has recently done. You might look at education and awareness about food and diet, as Jamie Oliver recently did very well, without recourse to dodgy pseudoscience or miracle pills.

You might even step away from obsessing over food – just for once – and look at parenting skills, teacher recruitment and retention, or social exclusion, or classroom size, or social inequality and the widening income gap. Or parenting programmes, as we said right at the beginning. But the media don't want stories like that. 'Pill solves complex social problem' feels much more like a news story than anything involving a boring parenting programme.

This is partly due to journalists' own sense of news value, but it's also a question of how stories are pushed. I've not met Hutchings *et al.*, the authors of the parenting study that kicked off this chapter – and I'm quite prepared to be told that they are in Soho House until 2 a.m. every night, schmoozing broadcast media journalists with champagne and nibbles – but in reality I suspect they are quiet, modest academics. Private companies, meanwhile, have top-dollar public-relations firepower, one single issue to promote, time to foster relationships with interested journalists, and a wily understanding of the desires of the public and the media, our collective hopes and consumer dreams.

The fish-oil story is by no means unique: repeatedly, in a bid to sell pills, people sell a wider explanatory framework, and as George Orwell first noted, the true genius in advertising is to sell you the solution *and* the problem. Pharmaceutical companies have worked hard, in their direct-to-consumer advertisements and their lobbying, to push the 'serotonin hypothesis' for depression, even though the scientific evidence for this theory is growing thinner every year; and the nutrition supplements industry, for its own market, promotes dietary deficiencies as a treatable cause for low mood (I myself do not have a miracle cure to offer, and repetitively enough, I think that the social causes of these problems are arguably more interesting – and possibly even more amenable to intervention).

These fish-oil stories were a classic example of a phenomenon more widely described as 'medicalisation', the expansion of the biomedical remit into domains where it may not be helpful or necessary. In the past, this has been portrayed as something that doctors inflict on a passive and unsuspecting world, an expansion of the medical empire: but in reality it seems that these reductionist biomedical stories can appeal to us all, because complex problems often have depressingly complex causes, and the solutions can be taxing and unsatisfactory.

In its most aggressive form, this process has been characterised as 'disease-mongering'. It can be seen throughout the world of quack cures – and being alive to it can be like having the scales removed from your eyes – but in big pharma the story goes like this: the low-hanging fruit of medical research has all been harvested, and the industry is rapidly running out of novel molecular entities. They registered fifty a year in the 1990s, but now it's down to twenty a year, and a lot of those are just copies. They are in trouble.

Because they cannot find *new treatments* for the diseases we already have, the pill companies instead invent *new diseases* for the treatments they already have. Recent favourites include Social Anxiety Disorder (a new use for SSRI drugs), Female Sexual Dysfunction (a new use for Viagra in women), night eating syndrome (SSRIs again) and so on: problems, in a real sense, but perhaps not necessarily the stuff of pills, and perhaps not best conceived of in reductionist biomedical terms. In fact, reframing intelligence, loss of libido, shyness and tiredness as medical pill problems could be considered crass, exploitative, and frankly disempowering.

These crude biomedical mechanisms may well enhance the placebo benefits from pills, but they are also seductive precisely because of what they edit out. In the media coverage around the rebranding of Viagra as a treatment for women in the early noughties, and the invention of the new disease Female Sexual Dysfunction, for example, it wasn't just the tablets that were being sold: it was the explanation.

Glossy magazines told stories about couples with relationship problems, who went to their GP, and the GP didn't understand their problem (because that is the first paragraph of any medical story in the media). Then they went to the specialist, and he didn't help either. But then they went to a private clinic. They did blood tests, hormone profiles, esoteric imaging studies of clitoral bloodflow, and they understood: the solution was in a pill, but that was only half the story. It was a mechanical problem. Rarely was there a mention of any other factors: that she was feeling tired from overwork, or he was exhausted from being a new father, and finding it hard to come to terms with the fact that his wife was now the mother of his children, and no longer the vixen he first snogged on the floor of the student union building to the sound of 'Don't You Want Me Baby?' by the Human League in 1983: no. Because we don't want to talk about these issues, any more than we want to talk about social inequality, the disintegration of local communities, the breakdown of the family, the impact of employment uncertainty, changing expectations and notions of personhood, or any of the other complex, difficult factors that play into the apparent rise of antisocial behaviour in schools.

But above all we should pay tribute to the genius of this huge fish-oil project, and every other nutritionist who has got their pills into the media, and into schools, because more than anything else, they have sold children, at the most impressionable time of their lives, one very compelling message: that you need to take pills to lead a healthy normal life, that a sensible diet and lifestyle are not enough in themselves, and that a pill can even make up for failings elsewhere. They have pushed their message directly into schools, into families, into the minds of their worried parents, and it is their intention that every child should understand that you need to eat several large, expensive, coloured capsules, six of them, three times a day, and this will improve vital but intangible qualities: concentration, behaviour and intelligence.

This is the greatest benefit to the pill industry, of every complexion. I would prefer fish-oil pills to ritalin, but fish-oil pills are being marketed at every child in the country, and they have undoubtedly won. Friends tell me that in some schools it is considered almost child neglect not to buy these capsules, and its impact on this generation of schoolchildren, reared on pills, will continue to bear rich fruit for all the industries, long after the fish-oil capsules have been forgotten.

Calming down: the apothecary industrial complex

Generating news coverage as a means of increasing brand awareness for a commercial product is a well-trodden path (and it also drives those contentless 'scientists have found the equation for …' stories that we shall see in a later chapter). PR companies even calculate something called 'advertising equivalents' for the exposure your brand gets for free, and in a period when more newsprint is generated by fewer journalists, it's inevitable that such shortcuts to colourful copy are welcomed by reporters. News and features stories about a product also carry far more weight in the public imagination than a paid advert, and are more likely to be read or watched.

But there is another, more subtle benefit to be had from editorial coverage of your pseudo-medical product: the claims that can be made in advertising and on packaging for food supplements and 'borderline medical products' are tightly regulated, but there are no such regulations covering the claims made by journalists.

This rather clever division of labour is one of the more interesting features of the alternative therapy industry. Take a moment to think of all the things that you have come to believe are true, or at least have heard claimed regularly, about different supplements: glucosamine can treat arthritis; antioxidants prevent cancer and heart disease; omega-3 fish oils improve intelligence. These claims are now common currency, a part of our culture as much as 'dock leaves for stinging nettles'; but you will rarely, if ever, see them being made explicitly on packaging or in advertising material.

Once you realise this, it makes the colour supplements a marginally more interesting read: the alternative therapy columnist will make a dramatic and scientifically untenable claim for glucosamine, stating that it will improve the joint pain of a reader who has written in; the pill company, meanwhile, will have a full-page advertisement for glucosamine, which merely states the dose and possibly makes a bland claim at the level of basic biology, rather than about clinical efficacy: 'Glucosamine is a known chemical constituent of cartilage.'

Sometimes the overlap is close enough to be amusing. Some examples are predictable. The vitamin pill magnate Patrick Holford, for example, makes sweeping and dramatic claims for all kinds of supplements in his 'Optimum Nutrition' books; yet these same claims are not to be found on the labels of his own-brand 'Optimum Nutrition' range of vitamin pills (which do feature, however, a photograph of his face).

Alternative health columnist Susan Clark – who has argued, among other things, that water has calories – is another shining example of this fine line that journalists can sometimes tread. She had a column in the *Sunday Times*, *Grazia* and the *Observer* for several years. In the course of these columns she recommended one company's products, 'Victoria Health', with notable frequency: once a month, regular as clockwork, by my reckoning. The papers deny that there was any impropriety, as does she, and I have no reason to doubt this. But she had done paid work for the company in the past, and she has now left her newspaper jobs to take up a full-time position at Victoria Health, writing their in-house magazine. (It's a scene that is eerily reminiscent of the well-known free flow of employees in

America between the pharmaceutical industry regulator and the boards of various pharmaceutical companies: in fact, at the risk of hammering a point home too hard, you will have noticed by now that I am telling the story of all the pill industries, using examples from mainstream media, and you couldn't put a cigarette paper between them.)

The Royal Pharmaceutical Society was expressing concern at these covert marketing strategies in the mainstream pharmaceutical industry as long ago as 1991: 'Barred from labelling products with detailed medicinal claims unless they submit to the licensing procedure,' it said, 'manufacturers and marketing companies are resorting to methods such as celebrity endorsements, free pseudomedical product literature, and press campaigns that have resulted in uncritically promotional features in large-circulation newspapers and magazines.'

Access to the unpoliced world of the media is well recognised as a major market advantage for Equazen, and it is an advantage that they press home hard. In the press release announcing the company's acquisition by the pharmaceutical company Galenica, they declared: 'Coverage of research showing the benefits of our Eye Q has appeared numerous times on national television and radio ... that is widely credited with being instrumental in the significant growth of the UK omega-3 sector since 2003.' To be honest, I would prefer to see a clearly labelled 'nonsense box' on all packaging and advertising, in which alternative therapy producers can freely make any claims they want to, instead of this misleading editorial coverage, because adverts are at least clearly labelled as such.

The wheels of time

Of course, these Durham trials are not the first time the world has seen such an extraordinary effort to promote one food supplement's powers through media stories on inaccessible research. David Horrobin was a 1980s pharmaceutical multi-millionaire – one of the richest men in Britain – and his Efamol food supplements empire (built, like Equazen's, on 'essential fatty acids') was worth an eye-watering £550 million at its peak. The efforts of his company went far further than anything one could find in the world of Equazen and Durham Council.

In 1984, staff at Horrobin's US distributors were found guilty in court of misbranding their food supplement as a drug; they were circumventing Food and Drug Administration regulations which forbade them from making unfounded claims for their supplement pills in advertising by engineering media coverage that treated them as if they had proven medical benefits. In the court case, paperwork was produced as evidence in which Horrobin explicitly said things like: 'Obviously you could not advertise [evening primrose oil] for these purposes but equally obviously there are ways of getting the information across …' Company memos described elaborate promotional schemes: planting articles on their research in the media, deploying researchers to make claims on their behalf, using radio phone-ins and the like.

In 2003 Horrobin's researcher Dr Goran Jamal was found guilty by the GMC of fraudulently concocting research data on trials which he had performed for Horrobin. He had been promised 0.5 per cent of the product's profits should it come to market (Horrobin was not responsible, but this is a fairly unusual payment arrangement which would rather dangle temptation in front of your eyes).

As with the fish-oil pills, Horrobin's products were always in the news, but it was difficult to get hold of the research data. In 1989 he published a famous meta-analysis of trials in a dermatology journal which found that his lead product, evening primrose oil, was effective in eczema. This meta-analysis excluded the one available large published trial (which was negative), but included the two oldest studies, and seven small positive studies sponsored by his own company (these were still unavailable at the last review I could find, in 2003).

In 1990 two academics had their review of the data binned by the journal after Horrobin's lawyers got involved. In 1995 the Department of Health commissioned a meta-analysis from a renowned epidemiologist. This included ten unpublished studies held by the company which was marketing evening primrose oil. The ensuing scene was only fully described by Professor Hywel Williams a decade later in an editorial for the *British Medical Journal*. The company erupted over a leak, and the Department of Health forced all authors and referees to sign written statements to reassure it. The academics were not allowed to publish their report. Alternative therapy, the people's medicine!

It has since been shown, following a wider but undisclosed review, that evening primrose oil is *not* effective in eczema, and it has lost its medicines licence. The case is still cited by leading figures in evidence-based medicine, such as Sir Iain Chalmers, founder of the Cochrane Collaboration, as an example of a pharmaceutical company refusing to release information on clinical trials to academics wishing to examine their claims.

David Horrobin, I feel duty bound to mention, is the father of the founding director of Equazen, Cathra Kelliher, *née* Horrobin, and her husband and co-director Adam Kelliher specifically cited him in interviews as a major influence on his business practices. I am not suggesting that their business practices are the same, but in my view the parallels – with inaccessible data, and research results that wind up being presented directly to the media – are striking.

In 2007 the GCSE results of the children in the Durham fish-oil year came in. This was an area of failing schools, receiving a huge amount of extra effort and input of all forms. The preceding year, with no fish oil, the results – the number of kids getting five GCSE grades A* to C – had improved by 5.5 per cent. After the fish-oil intervention the rate of improvement deteriorated notably, giving only a 3.5 per cent improvement. This was against a backdrop of a 2 per cent increase in GCSE scores nationally. You would have expected an improvement from a failing region whose schools were receiving a large amount of extra assistance and investment, and you might also remember, as we said, that GCSE results improve nationally every year. If anything, the pills seem to have been associated with a slowing of improvements.

Fish oils, meanwhile, are now the most popular food supplement product in the UK, with annual sales for that single product worth over £110 million a year. And the Kellihers recently sold Equazen to a major pharmaceutical corporation for an undisclosed sum. If you think I have been overly critical, I would invite you to notice that they win.

9

Professor Patrick Holford

Where do all these ideas about pills, nutritionists and fad diets come from? How are they generated, and propagated? While Gillian McKeith leads the theatrical battalions, Patrick Holford is a very different animal: he is the academic linchpin at the centre of the British nutritionism movement, and the founder of its most important educational establishment, the 'Institute for Optimum Nutrition'. This organisation has trained most of the people who describe themselves as 'nutrition therapists' in the UK.* Holford is, in many respects, the originator of their ideas, and the inspiration for their business practices.

Praise is heaped upon him in newspapers, where he is presented as an academic expert. His books are best-sellers, and he has written or collaborated on around forty. They have been translated into twenty languages, and have sold over a million

---

* 'Nutritionist', 'nutrition therapist', 'nutritional therapy consultant' and the many variations on this theme are not protected terms, unlike 'nurse', 'dietitian' or 'physiotherapist', so anyone can use them. Just to be clear, I'll say it again: anyone can declare themselves to be a nutritionist. After reading this book, you will know more about the appraisal of evidence than most, so in the manner of Spartacus I suggest you call yourself one too; and academics working in the field of nutrition will have to move on, because the word doesn't belong to them any more.

copies worldwide, to practitioners and the public. Some of his earlier works are charmingly fey, with one featuring a *Blue Peter*-ish 'dowsing kit' to help you diagnose nutritional deficiencies. The modern ones are drenched in scientific detail, and stylistically they exemplify what you might call 'referenciness': they have those nice little superscript numbers in the text, and lots of academic citations at the back.

Holford markets himself vigorously as a man of science, and he has recently been awarded a visiting professorship at the University of Teesside (on which more later). At various times he's had his own slot on daytime television, and hardly a week goes by without him appearing somewhere to talk about a recommendation, his latest 'experiment', or a 'study': one school experiment (with no control group) has been uncritically covered in two separate, dedicated programmes on *Tonight with Trevor MacDonald*, ITV's peak-hour investigative slot, and that sits alongside his other appearances on *This Morning*, BBC *Breakfast*, *Horizon*, BBC News, GMTV, *London Tonight*, Sky News, CBS News in America, *The Late Late Show* in Ireland, and many more. According to the British media establishment, Professor Patrick Holford is one of our leading public intellectuals: not a vitamin-pill salesman working in the $50-billion food-supplement industry – a fact which is very rarely mentioned, if ever – but an inspiring academic, embodying a diligent and visionary approach to scientific evidence. Let us see what calibre of work is required for journalists to accord you this level of authority in front of the nation.

AIDS, cancer and vitamin pills

I first became fully aware of Holford in a bookshop in Wales. It was a family holiday, I had nothing to write about, and it was New Year. Like a lifesaver, here was a copy of his *New Optimum Nutrition Bible*, the 500,000-copy best-seller. I seized it hungrily, and looked up the big killers. First I found a section heading which explains that 'people who take vitamin C live four times longer with cancer'. Excellent stuff.

I looked up AIDS (this is what I call 'the AIDS test'). Here is what I found on page 208: 'AZT, the first prescribable anti-HIV drug, is potentially harmful, and proving less effective than vitamin C.' Now, AIDS and cancer are very serious issues indeed. When you read a dramatic claim like Holford's, you might assume it's based on some kind of study, perhaps where people with AIDS were given vitamin C. There's a little super-script '23', referring you to a paper by someone called Jariwalla. With bated breath I grabbed a copy of this paper online.

The first thing I noticed was that this paper does not mention 'AZT'. It does not compare AZT with vitamin C. Nor docs it involve any human beings: it's a laboratory study, looking at some cells in a dish. Some vitamin C was squirted onto these cells, and a few complicated things were measured, like 'giant cell syncytia formation', which changed when there was lots of vitamin C swimming around. All well and good, but this laboratory-bench finding very clearly does not support the rather dramatic assertion that 'AZT, the first prescribable anti-HIV drug, is potentially harmful, and proving less effective than vitamin C.' In fact, it seems this is yet another example of that credulous extrapolation from preliminary laboratory data to clinical claim in real human beings that we have come to recog-nise as a hallmark of the 'nutritionist'.

But it gets more interesting. I casually pointed all this out in a newspaper article, and Dr Raxit Jariwalla himself appeared, writing a letter to defend his research paper against the accusa-tion that it was 'bad science'. This, to me, raised a fascinating question, and one which is at the core of this issue of 'referenci-ness'. Jariwalla's paper was a perfectly good one, and I have never said otherwise. It measured some complicated changes at a basic biological level in some cells in a dish on a lab bench, when they had lots of vitamin C squirted onto them. The methods and results were impeccably well described by Dr Jari-walla. I have no reason to doubt his clear description of what he did.

But the flaw comes in the interpretation. If Holford had said: 'Dr Raxit Jariwalla found that if you squirt vitamin C onto cells in a dish on a lab bench it seems to change the activity of some of their components,' and referenced the Jariwalla paper, that would have been fine. He didn't. He wrote: 'AZT, the first prescribable anti-HIV drug, is potentially harmful, and proving less effective than vitamin C.' The scientific research is one thing. What you claim it shows – your interpretation – is entirely separate. Holford's was preposterous overextrapolation.

I would have thought this was the point at which many people might have said: 'Yes, in retrospect, that was perhaps a little foolishly phrased.' But Professor Holford took a different tack. He has claimed that I had quoted him out of context (I did not: you can view the full page from his book online). He has claimed that he has corrected his book (you can read about this in a note at the back of the book you are holding). He has thrown around repeated accusations that I have only criticised him on this point because I am a pawn of big pharmaceutical corporations (I am not; in fact, bizarrely, I am one of their most vicious critics). Crucially, he suggested that I had focused on a trivial, isolated error.

A vaguely systematic review

The joy of a book is that you have plenty of space to play with. I have here my copy of *The New Optimum Nutrition Bible*. It's 'the book you have to read if you care about your health', according to the *Sunday Times* quote on the front cover. 'Invaluable', said the *Independent on Sunday*, and so on. I have decided to check every single reference, like a crazed stalker, and I will now dedicate the entire second half of this book to producing an annotated edition of Holford's weighty tome.

Only kidding.

There are 558 pages of plausible technical jargon in Holford's book, with complicated advice on what foods to eat, and which kinds of pills you should buy (in the 'resources' section it turns out that his own range of pills are 'the best'). For our sanity I have restricted our examination to one key section: the chapter where he explains why you should take supplements. Before we begin, we must be very clear: I am only interested in Professor Holford because he teaches the nutritionists who treat the nation, and because he has been given a professorship at Teesside University, with plans for him to teach students and supervise research. If Professor Patrick Holford is a man of science, and an academic, then we should treat him as one, with a scrupulously straight bat.

So, turning to Chapter 12, page 97 (I'm working from the 'completely revised and updated' 2004 edition, reprinted in 2007, if you'd like to follow the working at home), we can begin. You'll see that Holford is explaining the need to eat pills. This might be an apposite moment to mention that Professor Patrick Holford currently has his own range of best-selling pills, at least twenty different varieties, all featuring a photograph of his smiling face on the label. This range is available through the pill company BioCare, and his previous range, which you will see in older books, was sold by Higher Nature.*

My whole purpose in writing this book is to teach good science by examining the bad, so you will be pleased to hear that the very first claim Holford makes, in the very first paragraph of

---

* Oh, and he works for the pill company BioCare as their Head of Education and Science (I may have mentioned that the company is 30 per cent owned by pharmaceuticals company Elder). In fact, in many respects he has spent his whole life selling pills. His first job on leaving York with a 2:2 in psychology in the 1970s was as a vitamin-pill salesman for the pill company Higher Nature. He sold his most recent pill-selling company, Health Products for Life, for half a million pounds in 2007 to BioCare, and he now works for that company.

his key chapter, is a perfect example of a phenomenon we have already encountered: 'cherry-picking', or selecting the data that suits your case. He says there is a trial which shows that vitamin C will reduce the incidence of colds. But there is a gold-standard systematic review from Cochrane which brings together the evidence from all twenty-nine different trials on this subject, covering 11,000 participants in total, and concluded that there is no evidence that vitamin C prevents colds. Professor Holford doesn't give a reference for his single, unusual trial which contradicts the entire body of research meticulously summarised by Cochrane, but it doesn't matter: whatever it is, because it conflicts with the meta-analysis, we can be clear that it is cherry-picked.

Holford does give a reference, immediately afterwards, for a study where blood tests showed that seven out of ten subjects were deficient in vitamin B. There is an authoritative-looking superscript number in the text. Turning to the back of the book, we find that his reference for this study is a cassette you used to be able to buy from his own Institute for Optimum Nutrition (it's called *The Myth of the Balanced Diet*). We then have a twenty-five-year-old report from the Bateman Catering Organisation (who?), apparently with the wrong date; a paper on vitamin B12; some 'experiment' without a control reported in a 1987 ION pamphlet so obscure it's not even in the British Library (which has everything). Then there is a bland statement referenced to an article in the Institute for Optimum Nutrition's *Optimum Nutrition Magazine*, and an uncontroversial claim supported by a valid paper – the children of mothers who have taken folic acid during pregnancy have fewer birth defects, a well-established fact reflected in Department of Health guidelines – because there has to be a grain of common-sense truth in the spiel somewhere. Getting back to the action, we are told about a study on ninety schoolchildren who get 10 per cent higher IQ scores after taking a high-dose multivitamin pill, sadly without a reference, before a true gem: one paragraph with four references.

The first is to a study by the great Dr R.K. Chandra, a disgraced researcher whose papers have been discredited and retracted, who has been the subject of major articles on research fraud, including one by Dr Richard Smith in the *British Medical Journal* called 'Investigating the previous studies of a fraudulent author'. There is an entire three-part investigative documentary series on his worrying career made by Canada's CBC (you can watch it online), and at the conclusion of it he was, to all intents and purposes, in hiding in India. He has 120 different bank accounts in various tax havens, and he did, of course, patent his own multivitamin mixture, which he sells as an 'evidence-based' nutrition supplement for the elderly. The 'evidence' is largely derived from his own clinical trials.

In the name of scrupulous fairness, I am happy to clarify that much of this has come out since the first edition of Holford's book; but there had been serious questions about Chandra's research for some time, and nutrition academics were wary about citing it, simply because his findings seemed to be so incredibly positive. In 2002 he had resigned his university post and failed to answer questions about his papers or to produce his data when challenged by his employers. The paper that Patrick Holford is referring to was finally fully retracted in 2005. The next reference in this same paragraph of his book is to another Chandra paper. Two in a row is unfortunate.

Professor Holford follows this up with a reference to a review paper, claiming that thirty-seven out of thirty-eight studies looking at vitamin C (again) found it beneficial in treating (not *preventing*, as in his previous claim in the text above) the common cold. Thirty-seven out of thirty-eight sounds very compelling, but the definitive Cochrane review on the subject shows mixed evidence, and only a minor benefit at higher doses.

I hooked out the paper Professor Holford is referencing for this claim: it is a retrospective re-analysis of a review of trials, looking only at ones which were conducted before 1975. Holford's publishers describe this edition of the *Optimum Nutrition Bible* as 'COMPLETELY REVISED AND UPDATED TO INCLUDE THE LATEST CUTTING-EDGE RESEARCH'. It was published in the year in which I turned thirty, yet Holford's big reference for his claim about vitamin C and colds in this chapter is a paper which specifically only looks at trials from before I was one year old. Since this review was carried out, I have learnt to walk and talk, gone to primary school, upper school, three universities to do three degrees, worked as a doctor for a few years, got a column in the *Guardian*, and written a few hundred articles, not to mention this book. From my perspective, it is no exaggeration to say that 1975 is precisely a lifetime ago. As far as I am concerned, 1975 is not within living memory. Oh, and the paper Professor Holford references doesn't even seem to have thirty-eight trials in it, only fourteen. For a man who keeps going on about vitamin C, Professor Holford does seem to be a little unfamiliar with the contemporary literature. Perhaps if you are worried about your vitamin C intake you might want to buy some ImmuneC from the BioCare Holford range, at just £29.95 for 240 tablets, with his face on the bottle.*

---

* It's important to remember the difference between *preventing* colds, where the Cochrane review found no evidence of any benefit, and *treating* them, where Cochrane shows minor benefit at very high doses. There are, as you might imagine, cases of Holford eliding the two, and more recently, in a newsletter to his paying customers, he mangled the data in a way that might well startle the original authors. He took the modest 13.6 per cent reduction in cold duration for children taking high-dose vitamin C and claimed: 'This equates to up to a month less "cold" days per year for the average child.' For this claim to be true, the average child would be having more than two hundred days of cold symptoms a year. According to the review, children who had the highest number of colds might actually expect a reduction of four days per year. I could go on with the litany of errors in his mailouts, but there is a line between making a point and driving the reader away.

We'll go on. He cherry-picks the single most dramatically positive paper that I can find in the literature for vitamin E preventing heart attacks – a 75 per cent reduction, he claims. To give you a flavour of the references he doesn't tell you about, I have taken the trouble to go back in time and find the most up-to-date review reference, as the literature stood in 2003: a systematic review and meta-analysis, collected and published in the *Lancet*, which assessed all the papers published on the subject from decades previously, and found overall that there is no evidence that vitamin E is beneficial. You may be amused to know that the single positive trial referenced by Holford is not just the smallest, but also the briefest study in this review, by a wide margin. This is Professor Holford: pitched to teach and supervise at Teesside University, moulding young minds and preparing them for the rigours of academic life.

He goes on to make a string of extraordinary claims, and none has any reference whatsoever. Children with autism won't look you in the eye, but 'give these kids natural vitamin A and they look straight at you'. No reference. Then he makes four specific claims for vitamin B, claiming 'studies' but giving no references. I promise we're coming to a punchline. There's some more stuff about vitamin C; this time the reference is to Chandra (yet again).

Finally, on page 104, in a triumphant sprint finish, Professor Patrick Holford says that there are now oranges with no vitamin C in them at all. It's a popular myth among self-declared nutritionists (there is no other kind), and those who sell food-supplement pills, that our food is becoming less nutritious: in reality, many argue it may be more nutritious overall, because we eat more fresh and frozen fruit and veg, less tinned or dried stuff, and so they all get to the shops quicker, and thus with more nutrients (albeit at phenomenal cost to the environment). But Holford's vitamin claim is somewhat more extreme than the usual fare. These oranges are not just *less* nutritious: 'Yes, some supermarket oranges contain no vitamin C!'* Frightening stuff! Buy pills!

This chapter is not an isolated case. There is an entire website – Holfordwatch – devoted to examining his claims in eye-watering detail, with breathtaking clarity and obsessive referencing. There you will find many more errors repeated in Holford's other documents, and carefully dissected with wit and slightly frightening pedantry. It is a genuine joy to behold.

Professor?

A couple of interesting things arise from this realisation. Firstly, and importantly, since I am always keen to engage with people's ideas: how might you conduct a discussion with someone like Patrick Holford? He is constantly accusing others of 'not keeping up' with the literature. Anyone who doubts the value of his pills is a 'flat-earther', or a pawn of the pharmaceutical industry. He would pull out research claims and references. What would you do, given that you can't possibly read them on the spot? Being scrupulously polite, and yet firm, the only sensible answer, surely, would be to say: 'I'm not entirely sure I can accept your précis or your interpretation of that data without checking it myself.' This may not go down too well.

But the second point is more important. Holford has been appointed – as I might have mentioned briefly – a *professor* at Teesside. He brandishes this fact proudly in his press releases, as you would expect. And according to Teesside documents – there's a large set, obtained under the Freedom of Information Act, available online – the clear plan at his appointment was for Professor Holford to supervise research, and to teach university courses.

It is not a surprise to me that there are entrepreneurs and gurus – individuals – selling their pills and their ideas on the

---

* I would like to invite Professor Holford to send me a supermarket orange that has no vitamin C in it, via the publisher's address.

open market. In some strange sense I respect and admire their tenacity. But it strikes me that universities have a very different set of responsibilities, and in the field of nutrition there is a particular danger. Homeopathy degrees, at least, are transparent. The universities where it is taught are secretive and sheepish about their courses (perhaps because when the exam papers leak, it turns out that they're asking questions about 'miasma' – in 2008) but at least these degrees in alternative therapies are what they say on the tin.

The nutritionists' project is more interesting: this work takes the *form* of science – the language, the pills and the referenciness – making claims that superficially mirror the assertions made by academics in the field of nutrition, where there is much real science to be done. Occasionally there may be some good evidence for their assertions (although I can't imagine the point of taking health advice from someone who is only occasionally correct). But in reality the work of 'nutritionists' is often, as we have seen, rooted in New Age alternative therapy, and while reiki quantum energy healing is fairly clear about where it's coming from, nutritionists have adopted the cloak of scientific authority so plausibly, with a smattering of common-sense lifestyle advice and a few references, that most people have barely spotted the discipline for what it is. On very close questioning, some nutritionists will acknowledge that theirs is a 'complementary or alternative therapy', but the House of Lords inquiry into alternative medicines, for example, didn't even list it as one.

This proximity to real academic scientific work summons up sufficient paradoxes that it is reasonable to wonder what might happen in Teesside when Professor Holford begins to help shape young minds. In one room, we can only imagine, a full-time academic will teach that you should look at the *totality* of evidence rather than cherry-pick, that you cannot *overextrapolate* from preliminary lab data, that *referencing* should be accurate, and should reflect the content of the paper you are citing, and everything else that an academic department might teach about science and health. In another room, will there be Patrick Holford, exhibiting the scholarship we have already witnessed?

We can have one very direct insight into this clash from a recent Holford mailout. Periodically, inevitably, a large academic study will be published which finds no evidence of benefit from one of Patrick Holford's favoured pills. Often he will issue a confused and angry rebuttal, and these critiques are highly influential behind the scenes: snippets of them frequently appear in newspaper articles, and traces of their flawed logic emerge in discussions with nutritionists.

In one, for example, he attacked a meta-analysis of randomised controlled trials of antioxidants as being biased, because it excluded two trials he said were positive. In fact they were not trials, they were simply observational surveys, and so could never have been included. On the occasion we are interested in, Patrick Holford was angry about a meta-analysis on omega-3 fish-oil pills, co-authored by Professor Carolyn Summerbell: she holds the full-time academic chair in Nutrition at Teesside University, where she is also Assistant Dean of Research, with a long-standing track record of published academic research in the field of nutrition.

In this case, Holford seems quite simply not to understand the main results statistics in the paper's results blobbogram, which showed no benefit for the fish oils.* Furious at what he

---

* There is a more detailed explanation of his misunderstanding online, but for nerds, it seems he is amazed that several studies with a non-significant trend towards showing a benefit for fish-oil pills do not collectively add up to show a statistically significant benefit. This is in fact, as you know, rather commonplace. There are several other interesting criticisms to be made of the fish-oil research paper, as there always are of any research paper, but sadly this, from Holford, is not one of them.

thought he had found, Professor Holford then went on to accuse the authors of being pawns of the pharmaceutical industry (you may be spotting a pattern). 'What I find particularly deceptive is that this obvious skew is not even discussed in the research paper,' he says. 'It really makes me question the integrity of the authors and the journal.' He is talking here, remember, about the Professor of Nutrition at Teesside University and Assistant Dean of Research. Things then deteriorate further. 'Let's explore that for a minute with a "conspiracy theory" hat on. Last week pharmaceutical drug sales topped $600 billion. The number one best seller was Lipitor, a statin drug for lowering cholesterol. It brought in $12.9 billion …'

Let us be clear: there is no doubt that there are serious problems with the pharmaceutical industry – I should know, I teach both medical students and doctors on the subject, I write about it regularly in national newspapers, and I am about to walk you through their evils in the next chapter – but the answer to this problem is not bad scholarship, nor is it another substitute set of pills from a related industry. Enough.

How did Holford come to be appointed?

David Colquhoun is Emeritus Professor in Pharmacology at UCL, and runs a magnificently shouty science blog at dcscience.net. Concerned, he obtained the 'case' for Professor Holford's appointment using the Freedom of Information Act, and posted it online. There are some interesting finds. Firstly, Teesside accepts that this is an unusual case. It goes on to explain that Holford is director of the Food for the Brain Foundation, which will be donating funds for a PhD bursary, and that he could help in a university autism clinic.

I am not going to dwell on Holford's CV – because I want to stay focused on the science – but the one sent to Teesside makes a good starting point for a brief biography. It says that he was at York studying experimental psychology from 1973 to 1976, before studying in America under two researchers in mental health and nutrition (Carl Pfeiffer and Abram Hoffer), and then returning to the UK in 1980 to treat 'mental health patients with nutritional medicine'. In fact 1975 was the first year that York ran a degree in psychology. Holford actually attended from 1976 to 1979, and after getting a 2:2 degree he began his first job, working as a salesman for the supplement-pill company Higher Nature. So he was treating patients in 1980, one year out of this undergraduate degree. Not a problem. I'm just trying to get this clear in my mind.

He set up the Institute of Optimum Nutrition in 1982, and he was director until 1998: it must therefore have been a touching and unexpected tribute for Patrick in 1995 when the Institute conferred upon him a Diploma in Nutritional Therapy. Since he started but failed to complete his Mphil in nutrition at the University of Surrey twenty years ago, this Dip.ION from his own organisation remains his only qualification in nutrition.

I could go on, but I find it unseemly, and also these are dreary details. OK, one more, but you'll have to read the rest online:

> In 1986 he started researching the effects of nutrition on intelligence, collaborating with Gwillym Roberts, a headmaster and student at ION. This culminated in a randomised controlled trial testing the effects of improved nutrition on children's IQ – an experiment that was the subject of a *Horizon* documentary and published in the *Lancet* in 1988.

I have this *Lancet* paper in front of me. It does not feature Holford's name anywhere. Not as an author, and not even as an acknowledgement.

Let's get back to his science, post haste. Could Teesside have easily discovered that there were reasons to be concerned about Patrick Holford's take on science, without deploying any evidence, before they appointed him as a visiting professor? Yes. Simply by reading the brochures from his own company, Health

Products for Life. Among the many pills, for example, they might have found his promotion and endorsement of the QLink pendant, at just £69.99. The QLink is a device sold to protect you from terrifying invisible electromagnetic rays, which Holford is eager to talk about, and it cures many ills. According to Holford's catalogue:

> It needs no batteries as it is 'powered' by the wearer – the microchip is activated by a copper induction coil which picks up sufficient micro currents from your heart to power the pendant.

The manufacturers explain that the QLink corrects your 'energy frequencies'. It has been covered in praise by *The Times*, the *Daily Mail* and ITV's *London Today*, and it's easy to see why: it looks a bit like a digital memory card for a camera, with eight contact pads on the circuit board on the front, a hi-tech electronic component mounted in the centre, and a copper coil around the edge.

Last summer I bought one and took it to Camp Dorkbot, an annual festival for dorks held – in a joke taken too far – at a scout camp outside Dorking. Here, in the sunshine, some of the nation's more childish electronics geeks examined the QLink. We chucked probes at it, and tried to detect any 'frequencies' emitted, but no luck. Then we did what any dork does when presented with an interesting device: we broke it open. Drilling down, the first thing we came to was the circuit board. This, we noted with some amusement, was not in any sense connected to the copper coil, and therefore it is not powered by the coil, as claimed.

The eight copper pads did have some intriguing-looking circuit-board tracks coming out of them, but on close inspection these were connected to absolutely nothing. You might call them 'decorative'. I should mention, in the name of accuracy, that I'm not clear if I can call something a 'circuit board' when there is no 'circuit'.

Finally, there is a modern surface-mount electronic component soldered to the centre of the device, prominently on display through the clear plastic cover. It looks impressive, but whatever it is, it is connected to absolutely nothing. Close examination with a magnifying glass, and experiments with a multimeter and an oscilloscope, revealed that this component on the 'circuit board' was a zero-ohm resistor. This is a resistor that has no resistance: a bit of wire in a tiny box. It sounds like a useless component, but they're actually quite useful for bridging a gap between adjacent tracks on a circuit board. (I feel I should apologise for knowing that.)

Now, such a component is not cheap. We must assume that this is an extremely high-quality surface-mount resistor, manufactured to very high tolerances, well calibrated, and sourced in small quantities. You buy them on paper tape in seven-inch reels, each reel containing about 5,000 resistors, and you could easily pay as much as £0.005 for such a resistor. Sorry, I was being sarcastic. Zero ohm resistors are extremely cheap. That's the QLink pendant. No microchip. A coil connected to nothing. And a zero-ohm resistor, which costs half a penny, and is also connected to nothing.*

Teesside is only part of the story. Our main reason for showing an interest in Patrick Holford is his phenomenal influence on the nutritionist community in the UK. As I have mentioned, I have a huge respect for the people I am writing about in this book, and I am happy to flatter Holford by saying that the modern phenomenon of nutritionism which pervades

---

* I contacted qlinkworld.co.uk to discuss my findings. They kindly contacted the inventor, who informed me that they have always been clear that the QLink does not use electronics components 'in a conventional electronic way'. Apparently the energy pattern reprogramming work is done by some finely powdered crystal embedded in the resin. I think that means it's a New Age crystal pendant, in which case they could just say so.

every aspect of the media is in large part his doing, through the graduates of his phenomenally successful Institute for Optimum Nutrition, where he still teaches. This institute has trained the majority of self-styled nutrition therapists in the UK, including Vicki Edgson from *Diet Doctors* on Channel Five, and Ian Marber, owner of the extensive 'Food Doctor' product range. It has hundreds of students.

We've seen some examples of the standard of Holford's scholarship. What happens in his Institute? Are its students, we might wonder, being tutored in the academic ways of its founder?

As an outsider, it's hard to tell. If you visit the academic-sounding website, www.ion.ac.uk (registered before the current rules on academic .ac.uk web addresses), you won't find a list of academics on the staff, or research programmes in progress, in the way that you would, say, for the Institute for Cognitive Neurosciences in London. Nor will you find a list of academic publications. When I rang up the press office once to get one, I was told about some magazine articles, and then when I explained what I really meant, the press officer went away, and came back, and told me that ION was 'a research institute, so they don't have time for academic papers and stuff'.

Slowly, more so since Holford's departure as head (he still teaches there), the Institute of Optimum Nutrition has managed to squeeze some respectability out of its office space in south-west London. It has managed to get its diploma properly accredited, by the University of Luton, and it now counts as a 'foundation degree'. With one more year of study, if you can find anyone to take you – that is, the University of Luton – you can convert your ION diploma into a full BSc science degree.

If, in casual conversation with nutritionists, I question the standards of the ION, this accreditation is frequently raised, so we might look at it very briefly. Luton, previously the Luton

College of Higher Education, now the University of Bedford-
shire, was the subject of a special inspection by the Quality
Assurance Agency for Higher Education in 2005. The QAA is
there to 'safeguard the academic standards and quality of higher
education in the UK'.

When the QAA's report was published, the *Daily Telegraph* ran
an article about Luton titled: 'Is this the Worst University in
Britain?' The answer, I suspect, is yes. But of particular interest to
us is the way the report specifically singled out the slapdash
approach of the university towards validating external foundation
degrees (p.12, para 45 onwards). It states outright that in the view
of the auditing team, the expectations of the code of practice for
the assurance of academic quality and standards in higher education, specifically with regard to accrediting foundation degrees,
were simply not met. As they go – and I try not to read this kind of
document too often – this report is pretty full-on. If you look it up
online, I particularly recommend paragraphs 45 to 52.

At the very moment that this book was going to press, it tran-
spired that Professor Holford has resigned from his post as
visiting professor, citing reorganisation in the university. I have
time to add just one sentence, and it is this: It will not stop. He
is now looking for academic credibility elsewhere. The reality is
that this vast industry of nutritionism – and more importantly
than anything, this fascinating brand of scholarship – is now
penetrating, uncriticised, unnoticed, to the heart of our
academic system, because of our desperation to find easy
answers to big problems like obesity, our collective need for
quick fixes, the willingness of universities to work with industry
figures across the board, the admirable desire to give students
what they want, and the phenomenal mainstream credibility
that these pseudo-academic figures have attained, in a world
that has apparently forgotten the importance of critically
appraising all scientific claims.

There are other reasons why these ideas have gone unexamined. One is workload. Patrick Holford, for example, will occasionally respond on an issue of evidence, but often, it seems to me, by producing an even greater cloud of sciencey material: enough to shoo off many critics, perhaps, and certainly reassuring for the followers, but anybody daring to question must be ready to address a potentially exponential mass of content, both from Holford, and also from his extensive array of paid staff. It's extremely good fun.

There is also the PCC complaint against me (not upheld, and not even forwarded to the paper for comment), the lengthy legal letters, his claims that the *Guardian* has corrected articles critical of him (which it most certainly has not), and so on. He writes long letters, sent to huge numbers of people, accusing me and others critical of his work of some rather astonishing things. These claims appear in mail-outs to the customers of his pill shop, in letters to health charities I've never heard of, in emails to academics, and in vast web pages: endless thousands of words, mostly revolving around his repeated and rather incongruous claim that I am somehow in the pocket of big pharma. I am not, but I note with some delight that – as I may have mentioned – Patrick, who sold his own pill retail outfit for half a million pounds last year, now works for BioCare, which is 30 per cent owned by a pharmaceutical company.

I am therefore speaking directly to you now, Professor Patrick Holford. If we disagree on any point of scientific evidence, instead of this stuff about the pharmaceutical industry being out to get you, or a complaint, or a legal letter, instead of airily claiming that queries should be taken up with the scientist whose valid work you are – as I think I have shown – overinterpreting, instead of responding on a different question than the one which was posed, or any other form of theatrics, I would welcome professorial clarification, simply and clearly.

These are not complicated matters. Either it is acceptable to cherry-pick evidence on, say, vitamin E, or it is not. Either it is reasonable to extrapolate from lab data about cells in a dish to a clinical claim about people with AIDS, or it is not. Either an orange contains vitamin C, or it does not. And so on. Where you have made errors, perhaps you could simply acknowledge that, and correct them. I will always happily do so myself, and indeed have done so many times, on many issues, and felt no great loss of face.

I welcome other people challenging my ideas: it helps me to refine them.

10

Is Mainstream Medicine Evil?

So that was the alternative therapy industry. Its practitioners' claims are made directly to the public, so they have greater cultural currency; and while they use the same tricks of the trade as the pharmaceutical industry – as we have seen along the way – their strategies and errors are more transparent, so they make for a neat teaching tool. Now, once again, we should raise our game.

For this chapter you will also have to rise above your own narcissism. We will not be talking about the fact that your GP is sometimes rushed, or that your consultant was rude to you. We will not be talking about the fact that nobody could work out what was wrong with your knee, and we will not even be discussing the time that someone misdiagnosed your grandfather's cancer, and he suffered unnecessarily for months before a painful, bloody, undeserved and undignified death at the end of a productive and loving life.

Terrible things happen in medicine, when it goes right as well as when it goes wrong. Everybody agrees that we should work to minimise the errors, everybody agrees that doctors are sometimes terrible; if the subject fascinates you, then I encourage you to buy one of the libraries' worth of books on clinical governance. Doctors can be awful, and mistakes can be murderous, but the philosophy driving evidence-based medicine is not. How well does it work?

One thing you could measure is how much medical practice is evidence-based. This is not easy. From the state of current knowledge, around 13 per cent of all *treatments* have good evidence, and a further 21 per cent are likely to be beneficial. This sounds low, but it seems the more common treatments tend to have a better evidence base. Another way of measuring is to look at how much medical *activity* is evidence-based, taking consecutive patients, in a hospital outpatients clinic for example, looking at their diagnosis, what treatment they were given, and then looking at whether that treatment decision was based on evidence. These real-world studies give a more meaningful figure: lots were done in the 1990s, and it turns out, depending on speciality, that between 50 and 80 per cent of all medical activity is 'evidence-based'. It's still not great, and if you have any ideas on how to improve that, do please write about it.*

---

* I have argued on various occasions that, wherever possible, all treatment where there is uncertainty should be randomised, and in the NHS we are theoretically in a unique administrative position to be able to facilitate this, as a gift to the world. For all that you may worry about some of its decisions, the National Institute for Health and Clinical Excellence (NICE) has also had the clever idea of recommending that some treatments – where there is uncertainty about benefit – should only be funded by the NHS when given in the context of a trial (an 'Only in Research' approval). NICE is frequently criticised – it's a political body after all – for not recommending that the NHS funds apparently promising treatments. But acquiescing and funding a treatment when it is uncertain whether it does more good than harm is dangerous, as has been dramatically illustrated by various cases where promising treatments turned out ultimately to do more harm than good. We failed for decades to address uncertainties about the benefits of steroids for patients with brain injury: the CRASH trial showed that tens of thousands of people have died unnecessarily, because in fact they do more harm than good. In medicine, information saves lives.

Another good measure is what happens when things go wrong. The *British Medical Journal* is probably the most important medical journal in the UK. It recently announced the three most popular papers from its archive for 2005, according to an audit that assessed their use by readers, the number of times they were referenced by other academic papers, and so on. Each of these papers had a criticism of either a drug, a drug company or a medical activity as its central theme.

We can go through them briefly, so you can see for yourself how relevant the biggest papers from the most important medical journal are to your needs. The top-scoring paper was a case-control study which showed that patients had a higher risk of heart attack if they were taking the drugs rofecoxib (Vioxx), diclofenac or ibuprofen. At number two was a large meta-analysis of drug company data, which showed no evidence that SSRI antidepressants increase the risk of suicide, but found weak evidence for an increased risk of deliberate self-harm. In third place was a systematic review which showed an *association* between suicide attempts and the use of SSRIs, and critically highlighted some of the inadequacies around the reporting of suicides in clinical trials.

This is critical self-appraisal, and it is very healthy, but you will notice something else: all of those studies revolve around situations where drug companies withheld or distorted evidence. How does this happen?

The pharmaceutical industry

The tricks of the trade which we'll discuss in this chapter are probably more complicated than most of the other stuff in the book, because we will be making technical critiques of an industry's professional literature. Drug companies thankfully don't advertise direct to the public in the UK – in America you can find them advertising anxiety pills for your dog – so we are pulling apart the tricks they play on doctors, an audience which is in a slightly better position to call their bluff. This means that we'll first have to explain some background about how a drug comes to market. This is stuff that you will be taught at school when I become president of the one world government.

Understanding this process is important for one very clear reason. It seems to me that a lot of the stranger ideas people have about medicine derive from an emotional struggle with the very notion of a pharmaceutical industry. Whatever our political leanings, everyone is basically a socialist when it comes to healthcare: we all feel nervous about profit taking any role in the caring professions, but that feeling has nowhere to go. Big pharma is evil: I would agree with that premise. But because people don't understand exactly *how* big pharma is evil, their anger and indignation get diverted away from valid criticisms – its role in distorting data, for example, or withholding life-saving AIDS drugs from the developing world – and channelled into infantile fantasies. 'Big pharma is evil,' goes the line of reasoning, 'therefore homeopathy works and the MMR vaccine causes autism.' This is probably not helpful.

In the UK, the pharmaceutical industry has become the third most profitable activity after finance and – a surprise if you live here – tourism. We spend £7 billion a year on pharmaceutical drugs, and 80 per cent of that goes on patented drugs, medicines which were released in the last ten years. Globally, the industry is worth around £150 billion.

People come in many flavours, but all corporations have a duty to maximise their profits, and this often sits uncomfortably with the notion of caring for people. An extreme example comes with AIDS: as I mentioned in passing, drug companies explain that they cannot give AIDS drugs off licence to developing-world countries, because they need the money from sales for research and development. And yet, of the biggest US companies' $200 billion sales, they spend only 14 per cent on R&D, compared to 31 per cent on marketing and administration.

The companies also set their prices in ways you might judge to be exploitative. Once your drug comes out, you have around ten years 'on patent', as the only person who is allowed to make it. Loratadine, produced by Schering-Plough, is an effective antihistamine drug that does not cause the unpleasant antihistamine side-effect of drowsiness. It was a unique treatment for a while, and highly in demand. Before the patent ran out, the price of the drug was raised thirteen times in just five years, increasing by over 50 per cent. Some might regard this as profiteering.

But the pharmaceutical industry is also currently in trouble. The golden age of medicine has creaked to a halt, as we have said, and the number of new drugs, or 'new molecular entities', being registered has dwindled from fifty a year in the 1990s to about twenty now. At the same time, the number of 'me-too' drugs has risen, making up to half of all new drugs.

Me-too drugs are an inevitable function of the market: they are rough copies of drugs that already exist, made by another company, but are different enough for a manufacturer to be able to claim their own patent. They take huge effort to produce, and need to be tested (on human participants, with all the attendant risks) and trialled and refined and marketed just like a new drug. Sometimes they offer modest benefits (a more convenient dosing regime, for example), but for all the hard work they involve, they don't generally represent a significant breakthrough in human health. They are merely a breakthrough in making money. Where do all these drugs come from?

The journey of a drug

First of all, you need an idea for a drug. This can come from any number of places: a molecule in a plant; a receptor in the body that you think you can build a molecule to interface with; an old drug that you've tinkered with; and so on. This part of the story is extremely interesting, and I recommend doing a degree in it. When you think you have a molecule that might be a runner, you test it in animals, to see if it works for whatever you think it should do (and to see if it kills them, of course).

Then you do Phase I, or 'first in man', studies on a small number of brave, healthy young men who need money, firstly to see if it kills them, and also to measure basic things like how fast the drug is excreted from the body (this is the phase that went horribly wrong in the TGN1412 tests in 2006, where several young men were seriously injured). If this works, you move to a Phase II trial, in a couple of hundred people with the relevant illness, as a 'proof of concept', to work out the dose, and to get an idea if it is effective or not. A *lot* of drugs fail at this point, which is a shame, since this is no GCSE science project: bringing a drug to market costs around $500 million in total.

Then you do a Phase III trial, in hundreds or thousands of patients, randomised, blinded, comparing your drug against placebo or a comparable treatment, and collect much more data on efficacy and safety. You might need to do a few of these, and then you can apply for a licence to sell your drug. After it goes to market, you should be doing more trials, and other people will probably do trials and other studies on your drug too; and hopefully everyone will keep their eyes open for any previously unnoticed side-effects, ideally reporting them using the Yellow Card system (patients can use this too; in fact, please do. It's at http://yellowcard.mhra.gov.uk).

Doctors make their rational decision on whether they want to prescribe a drug based on how good it has been shown to be in trials, how bad the side-effects are, and sometimes cost. Ideally they will get their information on efficacy from studies published in peer-reviewed academic journals, or from other material like textbooks and review articles which are themselves based on primary research like trials. At worst, they will rely on the lies of drug reps and word of mouth.

But drug trials are expensive, so an astonishing 90 per cent of clinical drug trials, and 70 per cent of trials reported in major medical journals, are conducted or commissioned by the pharmaceutical industry. A key feature of science is that findings should be replicated, but if only one organisation is doing the funding, then this feature is lost.

It is tempting to blame the drug companies – although it seems to me that nations and civic organisations are equally at fault here for not coughing up – but wherever you draw your own moral line, the upshot is that drug companies have a huge influence over what gets researched, how it is researched, how the results are reported, how they are analysed, and how they are interpreted.

Sometimes whole areas can be orphaned because of a lack of money, and corporate interest. Homeopaths and vitamin pill quacks would tell you that their pills are good examples of this phenomenon. That is a moral affront to the better examples. There are conditions which affect a small number of people, like Creutzfeldt-Jakob disease and Wilson disease, but more chilling are the diseases which are neglected because they are only found in the developing world, like Chagas disease (which threatens a quarter of Latin America) and trypanosomiasis (300,000 cases a year, but in Africa). The Global Forum for Health Research estimates that only 10 per cent of the world's health burden receives 90 per cent of total biomedical research funding.

Often it is simply information that is missing, rather than some amazing new molecule. Eclampsia, say, is estimated to cause 50,000 deaths in pregnancy around the world each year, and the best treatment, by a huge margin, is cheap, unpatented, magnesium sulphate (high doses intravenously, that is, not some alternative medicine supplement, but also not the expensive anti-convulsants that were used for many decades). Although magnesium had been used to treat eclampsia since 1906, its position as the best treatment was only established a century later in 2002, with the help of the World Health Organisation, because there was no commercial interest in the research question: nobody has a patent on magnesium, and the majority of deaths from eclampsia are in the developing world. Millions of women have died of the condition since 1906, and many of those deaths were avoidable.

To an extent these are political and development issues, which we should leave for another day; and I have a promise to pay out on: you want to be able to take the skills you've learnt about levels of evidence and distortions of research, and understand how the pharmaceutical industry distorts data, and pulls the wool over our eyes. How would we go about proving this? Overall, it's true, drug company trials are much more likely to produce a positive outcome for their own drug. But to leave it there would be weak-minded.

What I'm about to tell you is what I teach medical students and doctors – here and there – in a lecture I rather childishly call 'drug company bullshit'. It is, in turn, what I was taught at medical school,* and I think the easiest way to understand the issue is to put yourself in the shoes of a big pharma researcher.

---

* In this subject, like many medics of my generation, I am indebted to the classic textbook *How to Read a Paper* by Professor Greenhalgh at UCL. It should be a best-seller. *Testing Treatments* by Imogen Evans, Hazel Thornton and Iain Chalmers is also a work of great genius, appropriate for a lay audience, and amazingly also free to download online from www.jameslindlibrary.org. For committed readers I recommend *Methodological Errors in Medical Research* by Bjorn Andersen. It's extremely long. The subtitle is 'An Incomplete Catalogue'.

You have a pill. It's OK, maybe not that brilliant, but a lot of money is riding on it. You need a positive result, but your audience aren't homeopaths, journalists or the public: they are doctors and academics, so they have been trained in spotting the obvious tricks, like 'no blinding', or 'inadequate randomisation'. Your sleights of hand will have to be much more elegant, much more subtle, but every bit as powerful.

What can you do?

Well, firstly, you could study it in winners. Different people respond differently to drugs: old people on lots of medications are often no-hopers, whereas younger people with just one problem are more likely to show an improvement. So only study your drug in the latter group. This will make your research much less applicable to the actual people that doctors are prescribing for, but hopefully they won't notice. This is so commonplace it is hardly worth giving an example.

Next up, you could compare your drug against a useless control. Many people would argue, for example, that you should *never* compare your drug against placebo, because it proves nothing of clinical value: in the real world, nobody cares if your drug is better than a sugar pill; they only care if it is better than the best currently available treatment. But you've already spent hundreds of millions of dollars bringing your drug to market, so stuff that: do lots of placebo controlled trials and make a big fuss about them, because they practically guarantee some positive data. Again, this is universal, because almost all drugs will be compared against placebo at some stage in their lives, and 'drug reps' – the people employed by big pharma to bamboozle doctors (many simply refuse to see them) – love the unambiguous positivity of the graphs these studies can produce.

Then things get more interesting. If you do have to compare your drug with one produced by a competitor – to save face, or because a regulator demands it – you could try a sneaky underhand trick: use an inadequate dose of the competing drug, so that patients on it don't do very well; or give a very high dose of the competing drug, so that patients experience lots of side-effects; or give the competing drug in the wrong way (perhaps orally when it should be intravenous, and hope most readers don't notice); or you could increase the dose of the competing drug much too quickly, so that the patients taking it get worse side-effects. Your drug will shine by comparison.

You might think no such thing could ever happen. If you follow the references in the back, you will find studies where patients were given really rather high doses of old-fashioned antipsychotic medication (which made the new-generation drugs look as if they were better in terms of side-effects), and studies with doses of SSRI antidepressants which some might consider unusual, to name just a couple of examples. I know. It's slightly incredible.

Of course, another trick you could pull with side-effects is simply not to ask about them; or rather – since you have to be sneaky in this field – you could be careful about how you ask. Here is an example. SSRI antidepressant drugs cause sexual side-effects fairly commonly, including anorgasmia. We should be clear (and I'm trying to phrase this as neutrally as possible): I *really* enjoy the sensation of orgasm. It's important to me, and everything I experience in the world tells me that this sensation is important to other people too. Wars have been fought, essentially, for the sensation of orgasm. There are evolutionary psychologists who would try to persuade you that the entirety of human culture and language is driven, in large part, by the pursuit of the sensation of orgasm. Losing it seems like an important side-effect to ask about.

And yet, various studies have shown that the reported prevalence of anorgasmia in patients taking SSRI drugs varies between 2 per cent and 73 per cent, depending primarily on how you ask: a casual, open-ended question about side-effects, for example, or a careful and detailed enquiry. One 3,000-subject review on SSRIs simply did not list any sexual side-effects on its twenty-three-item side-effect table. Twenty-three other things were more important, according to the researchers, than losing the sensation of orgasm. I have read them. They are not.

But back to the main outcomes. And here is a good trick: instead of a real-world outcome, like death or pain, you could always use a 'surrogate outcome', which is easier to attain. If your drug is supposed to reduce cholesterol and so prevent cardiac deaths, for example, don't measure cardiac deaths, measure reduced cholesterol instead. That's much easier to achieve than a reduction in cardiac deaths, and the trial will be cheaper and quicker to do, so your result will be cheaper *and* more positive. Result!

Now you've done your trial, and despite your best efforts things have come out negative. What can you do? Well, if your trial has been good overall, but has thrown out a few negative results, you could try an old trick: don't draw attention to the disappointing data by putting it on a graph. Mention it briefly in the text, and ignore it when drawing your conclusions. (I'm so good at this I scare myself. Comes from reading too many rubbish trials.)

If your results are completely negative, don't publish them at all, or publish them only after a long delay. This is exactly what the drug companies did with the data on SSRI antidepressants: they hid the data suggesting they might be dangerous, and they buried the data showing them to perform no better than placebo. If you're really clever, and have money to burn, then after you get disappointing data, you could do some more trials with the same protocol, in the hope that they will be positive: then try to bundle all the data up together, so that your negative data is swallowed up by some mediocre positive results.

Or you could get really serious, and start to manipulate the statistics. For two pages only, this book will now get quite nerdy.

I understand if you want to skip it, but know that it is here for the doctors who bought the book to laugh at homeopaths. Here are the classic tricks to play in your statistical analysis to make sure your trial has a positive result.

Ignore the protocol entirely

Always assume that any correlation *proves* causation. Throw all your data into a spreadsheet programme and report – as significant – any relationship between anything and everything if it helps your case. If you measure enough, some things are bound to be positive just by sheer luck.

Play with the baseline

Sometimes, when you start a trial, quite by chance the treatment group is already doing better than the placebo group. If so, then leave it like that. If, on the other hand, the placebo group is already doing better than the treatment group at the start, then adjust for the baseline in your analysis.

Ignore dropouts

People who drop out of trials are statistically much more likely to have done badly, and much more likely to have had side-effects. They will only make your drug look bad. So ignore them, make no attempt to chase them up, do not include them in your final analysis.

Clean up the data

Look at your graphs. There will be some anomalous 'outliers', or points which lie a long way from the others. If they are making your drug look bad, just delete them. But if they are helping your drug look good, even if they seem to be spurious results, leave them in.

'The best of five … no … seven … no … nine!'
If the difference between your drug and placebo becomes significant four and a half months into a six-month trial, stop the trial immediately and start writing up the results: things might get less impressive if you carry on. Alternatively, if at six months the results are 'nearly significant', extend the trial by another three months.

Torture the data
If your results are bad, ask the computer to go back and see if any particular subgroups behaved differently. You might find that your drug works very well in Chinese women aged fifty-two to sixty-one. 'Torture the data and it will confess to anything,' as they say at Guantanamo Bay.

Try every button on the computer
If you're really desperate, and analysing your data the way you planned does not give you the result you wanted, just run the figures through a wide selection of other statistical tests, even if they are entirely inappropriate, at random.

And when you're finished, the most important thing, of course, is to publish wisely. If you have a good trial, publish it in the biggest journal you can possibly manage. If you have a positive trial, but it was a completely unfair test, which will be obvious to everyone, then put it in an obscure journal (published, written and edited entirely by the industry): remember, the tricks we have just described hide nothing, and will be obvious to anyone who reads your paper, but only if they read it very attentively, so it's in your interest to make sure it isn't read beyond the abstract. Finally, if your finding is really embarrassing, hide it away somewhere and cite 'data on file'. Nobody will know the methods, and it will only be noticed if someone comes pestering you for the data to do a systematic review. Hopefully, that won't be for ages.

How can this be possible?

When I explain this abuse of research to friends from outside medicine and academia, they are rightly amazed. 'How can this be possible?' they say. Well, firstly, much bad research comes down to incompetence. Many of the methodological errors described above can come about by wishful thinking, as much as mendacity. But is it possible to prove foul play?

On an individual level, it is sometimes quite hard to show that a trial has been deliberately rigged to give the right answer for its sponsors. Overall, however, the picture emerges very clearly. The issue has been studied so frequently that in 2003 a systematic review found thirty separate studies looking at whether funding in various groups of trials affected the findings. Overall, studies funded by a pharmaceutical company were found to be four times more likely to give results that were favourable to the company than independent studies.

One review of bias tells a particularly Alice in Wonderland story. Fifty-six different trials comparing painkillers like ibuprofen, diclofenac and so on were found. People often invent new versions of these drugs in the hope that they might have fewer side-effects, or be stronger (or stay in patent and make money). In every single trial the sponsoring manufacturer's drug came out as better than, or equal to, the others in the trial. On not one occasion did the manufacturer's drug come out worse. Philosophers and mathematicians talk about 'transitivity': if A is better than B, and B is better than C, then C cannot be better than A. To put it bluntly, this review of fifty-six trials exposed a singular absurdity: all of these drugs were better than each other.

But there is a surprise waiting around the corner. Astonishingly, when the methodological flaws in studies are examined, it seems that industry-funded trials actually turn out to have *better* research methods, on average, than independent trials.

The most that could be pinned on the drug companies were some fairly trivial howlers: things like using inadequate doses of the competitor's drug (as we said above), or making claims in the conclusions section of the paper that exaggerated a positive finding. But these, at least, were transparent flaws: you only had to read the trial to see that the researchers had given a miserly dose of a painkiller; and you should always read the methods and results section of a trial to decide what its findings are, because the discussion and conclusion pages at the end are like the comment pages in a newspaper. They're not where you get your news from.

How can we explain, then, the apparent fact that industry funded trials are so often so glowing? How can all the drugs simultaneously be better than all of the others? The crucial kludge may happen after the trial is finished.

Publication bias and suppressing negative results

'Publication bias' is a very interesting and very human phenomenon. For a number of reasons, positive trials are more likely to get published than negative ones. It's easy enough to understand, if you put yourself in the shoes of the researcher. Firstly, when you get a negative result, it feels as if it's all been a bit of a waste of time. It's easy to convince yourself that you found nothing, when in fact you discovered a very useful piece of information: that the thing you were testing *doesn't work*.

Rightly or wrongly, finding out that something doesn't work probably isn't going to win you a Nobel Prize – there's no justice in the world – so you might feel demotivated about the project, or prioritise other projects ahead of writing up and submitting your negative finding to an academic journal, and so the data just sits, rotting, in your bottom drawer. Months pass. You get a new grant. The guilt niggles occasionally, but Monday's your day in clinic, so Tuesday's the beginning of the week really, and there's the departmental meeting on Wednesday, so Thursday's the only day you can get any proper work done, because Friday's your teaching day, and before you know it, a year has passed, your supervisor retires, the new guy doesn't even know the experiment ever happened, and the negative trial data is forgotten forever, unpublished. If you are smiling in recognition at this paragraph, then you are a very bad person.

Even if you do get around to writing up your negative finding, it's hardly news. You're probably not going to get it into a big-name journal, unless it was a massive trial on something everybody thought was really whizzbang until your negative trial came along and blew it out of the water, so as well as this being a good reason for you not bothering, it also means the whole process will be heinously delayed: it can take a year for some of the slacker journals to reject a paper. Every time you submit to a different journal you might have to re-format the references (hours of tedium). If you aim too high and get a few rejections, it could be years until your paper comes out, even if you are being diligent: that's years of people not knowing about your study.

Publication bias is common, and in some fields it is more rife than in others. In 1995, only 1 per cent of all articles published in alternative medicine journals gave a negative result. The most recent figure is 5 per cent negative. This is very, very low, although to be fair, it could be worse. A review in 1998 looked at the entire canon of Chinese medical research, and found that not one single negative trial had ever been published. Not one. You can see why I use CAM as a simple teaching tool for evidence-based medicine.

Generally the influence of publication bias is more subtle, and you can get a hint that publication bias exists in a field by doing something very clever called a funnel plot. This requires, only briefly, that you pay attention.

If there are lots of trials on a subject, then quite by chance they will all give slightly different answers, but you would

expect them all to cluster fairly equally around the true answer.
You would also expect that the bigger studies, with more participants in them, and with better methods, would be more closely
clustered around the correct answer than the smaller studies:
the smaller studies, meanwhile, will be all over the shop, unusually positive and negative at random, because in a study with,
say, twenty patients, you only need three freak results to send
the overall conclusions right off.

A funnel plot is a clever way of graphing this. You put the
effect (i.e., how effective the treatment is) on the x-axis, from
left to right. Then, on the y-axis (top-to-bottom, maths-skivers)
you put how big the trial was, or some other measure of how
accurate it was. If there is no publication bias, you should see a
nice inverted funnel: the big, accurate trials all cluster around
each other at the top of the funnel, and then as you go down the
funnel, the little, inaccurate trials gradually spread out to the
left and right, as they become more and more wildly inaccurate
– both positively and negatively.

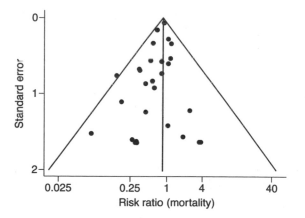

If there is publication bias, however, the results will be skewed.
The smaller, more rubbish *negative* trials seem to be missing,
because they were ignored – nobody had anything to lose by

letting these tiny, unimpressive trials sit in their bottom drawer – and so only the positive ones were published. Not only has publication bias been demonstrated in many fields of medicine, but a paper has even found evidence of publication bias in studies of publication bias. Here is the funnel plot for that paper. This is what passes for humour in the world of evidence-based medicine.

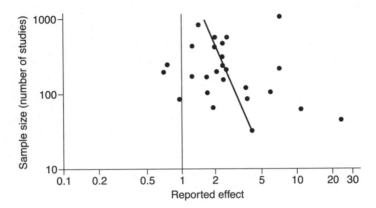

The most heinous recent case of publication bias has been in the area of SSRI antidepressant drugs, as has been shown in various papers. A group of academics published a paper in the *New England Journal of Medicine* at the beginning of 2008 which listed all the trials on SSRIs which had ever been formally registered with the FDA, and examined the same trials in the academic literature. Thirty-seven studies were assessed by the FDA as positive: with one exception, every single one of those positive trials was properly written up and published. Meanwhile, twenty-two studies that had negative or iffy results were simply not published at all, and eleven were written up and published in a way that described them as having a positive outcome.

This is more than cheeky. Doctors need reliable information if they are to make helpful and safe decisions about prescribing drugs to their patients. Depriving them of this information, and deceiving them, is a major moral crime. If I wasn't writing a light and humorous book about science right now, I would descend into gales of rage.

Duplicate publication

Drug companies can go one better than neglecting negative studies. Sometimes, when they get positive results, instead of just publishing them once, they publish them several times, in different places, in different forms, so that it looks as if there are lots of different positive trials. This is particularly easy if you've performed a large 'multicentre' trial, because you can publish overlapping bits and pieces from each centre separately, or in different permutations. It's also a very clever way of kludging the evidence, because it's almost impossible for the reader to spot.

A classic piece of detective work was performed in this area by a vigilant anaesthetist from Oxford called Martin Tramer, who was looking at the efficacy of a nausea drug called ondansetron. He noticed that lots of the data in a meta-analysis he was doing seemed to be replicated: the results for many individual patients had been written up several times, in slightly different forms, in apparently different studies, in different journals. Crucially, data which showed the drug in a better light were more likely to be duplicated than the data which showed it to be less impressive, and overall this led to a 23 per cent overestimate of the drug's efficacy.

Hiding harm

That's how drug companies dress up the positive results. What about the darker, more headline-grabbing side, where they hide the serious harms?

Side-effects are a fact of life: they need to be accepted, managed in the context of benefits, and carefully monitored, because the unintended consequences of interventions can be extremely serious. The stories that grab the headlines are ones where there is foul play, or a cover-up, but in fact important findings can also be missed for much more innocent reasons, like the normal human processes of accidental neglect in publication bias, or because the worrying findings are buried from view in the noise of the data.

Anti-arrhythmic drugs are an interesting example. People who have heart attacks get irregular heart rhythms fairly commonly (because bits of the timekeeping apparatus in the heart have been damaged), and they also commonly die from them. Anti-arrhythmic drugs are used to treat and prevent irregular rhythms in people who have them. Why not, thought doctors, just give them to everyone who has had a heart attack? It made sense on paper, they seemed safe, and nobody knew at the time that they would actually increase the risk of death in this group – because that didn't make sense from the theory (like with antioxidants). But they do, and at the peak of their use in the 1980s, anti-arrhythmic drugs were causing comparable numbers of deaths to the total number of Americans who died in the Vietnam war. Information that could have helped to avert this disaster was sitting, tragically, in a bottom drawer, as a researcher later explained:

> When we carried out our study in 1980 we thought that the increased death rate ... was an effect of chance ...The development of [the drug] was abandoned for commercial reasons, and so this study was therefore never published; it is now a good example of 'publication bias'. The results described here ... might have provided an early warning of trouble ahead.

That was neglect, and wishful thinking. But sometimes it seems that dangerous effects from drugs can be either deliberately downplayed or, worse than that, simply not published.

There has been a string of major scandals from the pharmaceutical industry recently, in which it seems that evidence of harm for drugs including Vioxx and the SSRI antidepressants has gone missing in action. It didn't take long for the truth to out, and anybody who claims that these issues have been brushed under the medical carpet is simply ignorant. They were dealt with, you'll remember, in the three highest-ranking papers in the *BMJ*'s archive. They are worth looking at again, in more detail.

Vioxx

Vioxx was a painkiller developed by the company Merck and approved by the American FDA in 1999. Many painkillers can cause gut problems – ulcers and more – and the hope was that this new drug might not have such side-effects. This was examined in a trial called VIGOR, comparing Vioxx with an older drug, naproxen: and a lot of money was riding on the outcome. The trial had mixed results. Vioxx was no better at relieving the symptoms of rheumatoid arthritis, but it did halve the risk of gastrointestinal events, which was excellent news. But an increased risk of heart attacks was also found.

When the VIGOR trial was published, however, this cardiovascular risk was hard to see. There was an 'interim analysis' for heart attacks and ulcers, where ulcers were counted for longer than heart attacks. It wasn't described in the publication, and it overstated the advantage of Vioxx regarding ulcers, while understating the increased risk of heart attacks. 'This untenable feature of trial design,' said a swingeing and unusually critical editorial in the *New England Journal of Medicine*, 'which inevitably skewed the results, was not disclosed to the editors or the academic authors of the study.' Was it a problem? Yes. For one thing, three additional myocardial infarctions occurred in the Vioxx group in the month after they stopped counting, while none occurred in the naproxen control group.

An internal memo from Edward Scolnick, the company's chief scientist, shows that the company knew about this cardiovascular risk ('It is a shame but it is a low incidence and it is mechanism based as we worried it was'). The *New England Journal of Medicine* was not impressed, publishing a pair of spectacularly critical editorials.

The worrying excess of heart attacks was only really picked up by people examining the FDA data, something that doctors tend – of course – not to do, as they read academic journal articles at best. In an attempt to explain the moderate extra risk of heart attacks that *could* be seen in the final paper, the authors proposed something called 'the naproxen hypothesis': Vioxx wasn't causing heart attacks, they suggested, but naproxen was preventing them. There is no accepted evidence that naproxen has a strong protective effect against heart attacks.

The internal memo, discussed at length in the coverage of the case, suggested that the company was concerned at the time. Eventually more evidence of harm emerged. Vioxx was taken off the market in 2004; but analysts from the FDA estimated that it caused between 88,000 and 139,000 heart attacks, 30 to 40 per cent of which were probably fatal, in its five years on the market. It's hard to be sure if that figure is reliable, but looking at the pattern of how the information came out, it's certainly felt, very widely, that both Merck and the FDA could have done much more to mitigate the damage done over the many years of this drug's lifespan, after the concerns were apparent to them. Data in medicine is important: it means lives. Merck has not admitted liability, and has proposed a $4.85 billion settlement in the US.

Authors forbidden to publish data

This all seems pretty bad. Which researchers are doing it, and why can't we stop them? Some, of course, are mendacious. But many have been bullied or pressured not to reveal information about the trials they have performed, funded by the pharmaceutical industry.

Here are two extreme examples of what is, tragically, a fairly common phenomenon. In 2000, a US company filed a claim against both the lead investigators and their universities in an attempt to block publication of a study on an HIV vaccine that found the product was no better than placebo. The investigators felt they had to put patients before the product. The company felt otherwise. The results were published in JAMA that year.

In the second example, Nancy Olivieri, director of the Toronto Haemoglobinopathies Programme, was conducting a clinical trial on deferiprone, a drug which removes excess iron from the bodies of patients who become iron-overloaded after many blood transfusions. She was concerned when she saw that iron concentrations in the liver seemed to be poorly controlled in some of the patients, exceeding the safety threshold for increased risk of cardiac disease and early death. More extended studies suggested that deferiprone might accelerate the development of hepatic fibrosis.

The drug company, Apotex, threatened Olivieri, repeatedly and in writing, that if she published her findings and concerns they would take legal action against her. With great courage – and, shamefully, without the support of her university – Olivieri presented her findings at several scientific meetings and in academic journals. She believed she had a duty to disclose her concerns, regardless of the personal consequences. It should never have been necessary for her to need to make that decision.

The single cheap solution that will solve all of the problems in the entire world

What's truly extraordinary is that almost all of these problems – the suppression of negative results, data dredging, hiding unhelpful data, and more – could largely be solved with one very simple intervention that would cost almost nothing: a clinical trials register, public, open, and properly enforced. This is how it would work. You're a drug company. Before you even start your study, you publish the 'protocol' for it, the methods section of the paper, somewhere public. This means that everyone can see what you're going to do in your trial, what you're going to measure, how, in how many people, and so on, *before you start.*

The problems of publication bias, duplicate publication and hidden data on side-effects – which all cause unnecessary death and suffering – would be eradicated overnight, in one fell swoop. If you registered a trial, and conducted it, but it didn't appear in the literature, it would stick out like a sore thumb. Everyone, basically, would assume you had something to hide, because you probably would. There are trials registers at present, but they are a mess.

How much of a mess is illustrated by this last drug company ruse: 'moving the goalposts'. In 2002 Merck and Schering Plough began a trial to look at Ezetimibe, a drug to reduce cholesterol. They started out saying they were going to measure one thing as their test of whether the drug worked, but then announced, after the results were in, that they were going to count something else as the real test instead. This was spotted, and they were publicly rapped. Why? Because if you measure lots of things (as they did), some might be positive simply by chance. You cannot find your starting hypothesis in your final results. It makes the stats go all wonky.

Adverts

> 'Clomicalm tablets are the only medication approved for the treatment of separation anxiety in dogs.'

There are currently no direct-to-consumer drug adverts in Britain, which is a shame, because the ones in America are properly bizarre, especially the TV ones. Your life is in disarray, your restless legs/migraine/cholesterol have taken over, all is panic, there is no sense anywhere. Then, when you take the right pill, suddenly the screen brightens up into a warm yellow, granny's laughing, the kids are laughing, the dog's tail is wagging, some nauseating child is playing with the hose on the lawn, spraying a rainbow of water into the sunshine whilst absolutely laughing his head off as all your relationships suddenly become successful again. Life is good.

Patients are so much more easily led than doctors by drug company advertising that the budget for direct-to-consumer advertising in America has risen twice as fast as the budget for addressing doctors directly. These adverts have been closely studied by medical academic researchers, and have been repeatedly shown to increase patients' requests for the advertised drugs, as well as doctors' prescriptions for them. Even adverts 'raising awareness of a condition' under tighter Canadian regulations have been shown to double demand for a specific drug to treat that condition.

This is why drug companies are keen to sponsor patient groups, or to exploit the media for their campaigns, as has been seen recently in the news stories singing the praises of the breast cancer drug Herceptin, or Alzheimer's drugs of borderline efficacy.

These advocacy groups demand vociferously in the media that the companies' drugs be funded by the NHS. I know people associated with these patient advocacy groups – academics – who have spoken out and tried to change their stance, without success: because in the case of the British Alzheimer's campaign in particular, it struck many people that the demands were rather one-sided. The National Institute for Clinical Excellence (NICE) concluded that it couldn't justify paying for Alzheimer's drugs, partly because the evidence for their efficacy was weak, and often looked only at soft, surrogate outcomes. The evidence is indeed weak, because the drug companies have failed to subject their medications to sufficiently rigorous testing on real-world outcomes, rigorous testing that would be much less guaranteed to produce a positive result. Does the Alzheimer's Society challenge the manufacturers to do better research? Do its members walk around with large placards campaigning against 'surrogate outcomes in drugs research', demanding 'More Fair Tests'? No.

Oh God. Everybody's bad. How did things get so awful?

11

How the Media Promote the Public Misunderstanding of Science

We need to make some sense of all this, and appreciate just how deep into our culture the misunderstandings and misrepresentations of science go. If I am known at all, it is for dismantling foolish media stories about science: it is the bulk of my work, my *oeuvre*, and I am slightly ashamed to say that I have over five hundred stories to choose from, in illustrating the points I intend to make here. You may well count this as obsessional.

We have covered many of the themes elsewhere: the seductive march to medicalise everyday life; the fantasies about pills, mainstream and quack; and the ludicrous health claims about food, where journalists are every bit as guilty as nutritionists. But here I want to focus on the stories that can tell us about the way science is perceived, and the repetitive, structural patterns in how we have been misled.

My basic hypothesis is this: the people who run the media are humanities graduates with little understanding of science, who wear their ignorance as a badge of honour. Secretly, deep down, perhaps they resent the fact that they have denied themselves access to the most significant developments in the history of Western thought from the past two hundred years; but there is an attack implicit in all media coverage of science: in their choice of stories, and the way they cover them, the media create a parody of science. On this template, science is portrayed as groundless, incomprehensible, didactic truth statements from scientists, who themselves are socially powerful, arbitrary, unelected authority figures. They are detached from reality; they do work that is either wacky or dangerous, but either way, everything in science is tenuous, contradictory, probably going to change soon and, most ridiculously, 'hard to understand'. Having created this parody, the commentariat then attack it, as if they were genuinely critiquing what science is all about.

Science stories generally fall into one of three categories: the wacky stories, the 'breakthrough' stories, and the 'scare' stories. Each undermines and distorts science in its own idiosyncratic way. We'll do them in order.

Wacky stories – money for nothing
If you want to get your research in the media, throw away the autoclave, abandon the pipette, delete your copy of *Stata*, and sell your soul to a PR company.

At Birmingham University there is a man called Dr Kevin Warwick, and he has been a fountain of eye-catching stories for some time. He puts a chip from a wireless ID card in his arm, then shows journalists how he can open doors in his department using it. 'I am a cyborg,' he announces, 'a melding of man and machine,'* and the media are duly impressed. A favourite research story from his lab – although it's never been published in any kind of academic journal, of course – purported to show that watching *Richard and Judy* improves children's IQ test performance much more effectively than all kinds of other

---

* This is a paraphrase, but it's not entirely inaccurate.

things you might expect to do so, like, say, some exercise, or drinking some coffee.

This was not a peripheral funny: it was a news story, and unlike most genuine science stories, it produced an editorial leader in the *Independent*. I don't have to scratch around to find more examples: there are five hundred to choose from, as I've said. 'Infidelity is genetic,' say scientists. 'Electricity allergy real,' says researcher. 'In the future, all men will have big willies,' says an evolutionary biologist from LSE.

These stories are empty, wacky filler, masquerading as science, and they reach their purest form in stories where scientists have 'found' the formula for something. How wacky those boffins are. Recently you may have enjoyed the perfect way to eat ice cream $(A \times Tp \times Tm/Ft \times At + V \times LT \times Sp \times W/Tt = 3d20)$, the perfect TV sitcom $(C = 3d[(R \times D) + V] \times F/A + S$, according to the *Telegraph*), the perfect boiled egg (*Daily Mail*), the perfect joke (the *Telegraph* again), and the most depressing day of the year $([W + (D-d)] \times TQ \ M \times NA$, in almost every newspaper in the world). I could go on.

These stories are invariably written up by science correspondents, and hotly followed – to universal approbation – by comment pieces from humanities graduates on how bonkers and irrelevant scientists are, because from the bunker-like mentality of my 'parody' hypothesis, that is the appeal of these stories: they play on the public's view of science as irrelevant, peripheral boffinry.

They are also there to make money, to promote products, and to fill pages cheaply, with a minimum of journalistic effort. Let's take some of the most prominent examples. Dr Cliff Arnall is the king of the equation story, and his recent output includes the formulae for the most miserable day of the year, the happiest day of the year, the perfect long weekend and many, many more. According to the BBC he is 'Professor Arnall'; usually he is 'Dr Cliff Arnall of Cardiff University'. In reality he's a private entrepreneur running confidence-building and stress-management courses, who has done a bit of part-time instructing at Cardiff University. The university's press office, however, are keen to put him in their monthly media-monitoring success reports. This is how low we have sunk.

Perhaps you nurture fond hopes for these formulae – perhaps you think they make science 'relevant' and 'fun', a bit like Christian rock. But you should know that they come from PR companies, often fully-formed and ready to have a scientist's name attached to them. In fact PR companies are very open to their customers about this practice: it is referred to as 'advertising equivalent exposure', whereby a 'news' story is put out which can be attached to a client's name.

Cliff Arnall's formula to identify the most miserable day of the year has now become an annual media stalwart. It was sponsored by Sky Travel, and appeared in January, the perfect time to book a holiday. His 'happiest day of the year' formula appears in June – it received yet another outing in the *Telegraph* and the *Mail* in 2008 – and was sponsored by Wall's ice cream. Professor Cary Cooper's formula to grade sporting triumphs was sponsored by Tesco. The equation for the beer-goggle effect, whereby ladies become more attractive after some ale, was produced by Dr Nathan Efron, Professor of Clinical Optometry at the University of Manchester, and sponsored by the optical products manufacturer Bausch & Lomb; the formula for the perfect penalty kick, by Dr David Lewis of Liverpool John Moores, was sponsored by Ladbrokes; the formula for the perfect way to pull a Christmas cracker, by Dr Paul Stevenson of the University of Surrey, was commissioned by Tesco; the formula for the perfect beach, by Dr Dimitrios Buhalis of the University of Surrey, sponsored by travel firm Opodo. These are people from proper universities, putting their names to advertising equivalent exposure for PR companies.

I know how Dr Arnall is paid, because when I wrote critically in the newspaper about his endless equations stories just before Christmas, he sent me this genuinely charming email:

> Further to your mentioning my name in conjunction with 'Walls' I just received a cheque from them. Cheers and season's greetings, Cliff Arnall.

It's not a scandal: it's just stupid. These stories are not informative. They are promotional activity masquerading as news. They play – rather cynically – on the fact that most news editors wouldn't know a science story if it danced naked in front of them. They play on journalists being short of time but still needing to fill pages, as more words are written by fewer reporters. It is, in fact, a perfect example of what investigative journalist Nick Davies has described as *Churnalism*, the uncritical rehashing of press releases into content, and in some respects this is merely a microcosm of a much wider problem that generalises to all areas of journalism. Research conducted at Cardiff University in 2007 showed that 80 per cent of all broadsheet news stories were 'wholly, mainly or partially constructed from second-hand material, provided by news agencies and by the public relations industry'.

It strikes me that you can read press releases on the internet, without paying for them in newsagents.

'All men will have big willies'
For all that they are foolish PR slop, these stories can have phenomenal penetrance. Those willies can be found in the *Sun*'s headline for a story on a radical new 'Evolution Report' by Dr Oliver Curry, 'evolution theorist' from the Darwin@LSE research centre. The story is a classic of the genre.

> By the year 3000, the average human will be 6½ft tall, have coffee-coloured skin and live for 120 years, new research predicts. And the good news does not end there. Blokes will be chuffed to learn their willies will get bigger – and women's boobs will become more pert.

This was presented as important 'new research' in almost every British newspaper. In fact it was just a fanciful essay from a political theorist at LSE. Did it hold water, even on its own terms?

No. Firstly, Dr Oliver Curry seems to think that geographical and social mobility are new things, and that they will produce uniformly coffee-coloured humans in 1,000 years. Oliver has perhaps not been to Brazil, where black Africans, white Europeans and Native Americans have been having children together for many centuries. The Brazilians have not gone coffee-coloured: in fact they still show a wide range of skin pigmentation, from black to tan. Studies of skin pigmentation (some specifically performed in Brazil) show that skin pigmentation seems not to be related to the extent of your African heritage, and suggest that colour may be coded for by a fairly small number of genes, and probably doesn't blend and even out as Oliver suggests.

What about his other ideas? He theorised that ultimately, through extreme socio-economic divisions in society, humans will divide into two species: one tall, thin symmetrical, clean, healthy, intelligent and creative; the other short, stocky, asymmetrical, grubby, unhealthy and not as bright. Much like the peace-loving Eloi and the cannibalistic Morlocks in H.G. Wells' *The Time Machine.*

Evolutionary theory is probably one of the top three most important ideas of our time, and it seems a shame to get it wrong. This ridiculous set of claims was covered in every British newspaper as a news story, but none of them thought to mention that dividing into species, as Curry thinks we will do, usually requires some fairly strong pressures, like, say, geographical divisions. The Tasmanian Aboriginals, for example, who had been isolated for 10,000 years, were still able to have children with other humans from outside. 'Sympatric speciation', a division into species where the two groups live in the same place, divided only by socioeconomic factors, as Curry is proposing, is even tougher. For a while, many scientists didn't think it happened at all. It would require that these divides were absolute, although history shows that attractive impoverished females and wealthy ugly men can be remarkably resourceful in love.

I could go on – the full press release is at badscience.net for your amusement. But the trivial problems in this trivial essay are not the issue: what's odd is how it became a 'boffins today said' science story all over the media, with the BBC, the *Telegraph*, the *Sun*, the *Scotsman*, *Metro* and many more lapping it up without criticism.

How does this happen? By now you don't need me to tell you that the 'research' – or 'essay' – was paid for by Bravo, a bikini-and-fast-car 'men's TV channel' which was celebrating its twenty-first year in operation. (In the week of Dr Curry's important science essay, just to give you a flavour of the channel, you could catch the movie classic *Temptations*: 'When a group of farm workers find that the bank intends to foreclose on their property, they console each other with a succession of steamy romps.' This might go some way to explaining the 'pert breasts' angle of his 'new research'.)

I spoke to friends on various newspapers, proper science reporters who told me they had stand-up rows with their newsdesks, trying to explain that this was not a science news story. But if they refused to write it, some other journalist would – you will often find that the worst science stories are written by consumer correspondents, or news generalists – and if I can borrow a concept from evolutionary theory myself, the selection pressure on employees in national newspapers is for journalists who compliantly and swiftly write up commercial puff nonsense as 'science news'.

One thing that fascinates me is this: Dr Curry is a proper academic (although a political theorist, not a scientist). I'm not seeking to rubbish his career. I'm sure he's done lots of stimulating work, but in all likelihood nothing he will ever do in his profession as a relatively accomplished academic at a leading Russell Group university will ever generate as much media coverage – or have as much cultural penetrance – as this childish, lucrative, fanciful, wrong essay, which explains nothing to anybody. Isn't life strange?

'Jessica Alba has the perfect wiggle, study says'

That's a headline from the *Daily Telegraph*, over a story that got picked up by Fox News, no less, and in both cases it was accompanied by compelling imagery of some very hot totty. This is the last wacky story we'll do, and I'm only including this one because it features some very fearless undercover work.

'Jessica Alba, the film actress, has the ultimate sexy strut, according to a team of Cambridge mathematicians.' This important study was the work of a team – apparently – headed by Professor Richard Weber of Cambridge University. I was particularly delighted to see it finally appear in print since, in the name of research, I had discussed prostituting my own reputation for it with Clarion, the PR company responsible, six months earlier, and there's nothing like watching flowers bloom.

Here is their opening email:

We are conducting a survey into the celebrity top ten sexiest walks for my client Veet (hair removal cream) and we would like to back up our survey with an equation from an expert to work out which celebrity has the sexiest walk, with theory behind it.

We would like help from a doctor of psychology or someone similar who can come up with equations to back up our findings, as we feel that having an expert comment and an equation will give the story more weight.

It got them, as we have seen, onto the news pages of the *Daily Telegraph*.

I replied immediately. 'Are there any factors you would particularly like to have in the equation?' I asked. 'Something sexual, perhaps?' 'Hi Dr Ben,' replied Kiren. 'We would really like the factors of the equation to include the thigh to calf ratio, the shape of the leg, the look of the skin and the wiggle (swing) of the hips … There is a fee of £500 which we would pay for your services.'

There was survey data too. 'We haven't conducted the survey yet,' Kiren told me, 'but we know what results we want to achieve.' That's the spirit! 'We want Beyoncé to come out on top followed by other celebrities with curvy legs such as J-Lo and Kylie and celebrities like Kate Moss and Amy Winehouse to be at the bottom e.g. – skinny and pale unshapely legs are not as sexy.' The survey, it turned out, was an internal email sent around the company. I rejected their kind offer, and waited. Professor Richard Weber did not. He regrets it. When the story came out, I emailed him, and it turned out that things were even more absurd than was necessary. Even after rigging their survey, they had to re-rig it:

The Clarion press release was not approved by me and is factually incorrect and misleading in suggesting there has been any serious attempt to do serious mathematics here. No 'team of Cambridge mathematicians' has been involved. Clarion asked me to help by analysing survey data from eight hundred men in which they were asked to rank ten celebrities for 'sexiness of walk'. And Jessica Alba did not come top. She came seventh.

Are these stories so bad? They are certainly pointless, and reflect a kind of contempt for science. They are merely PR promotional pieces for the companies which plant them, but it's telling that they know exactly where newspapers' weaknesses lie: as we shall see, bogus survey data is a hot ticket in the media.

And did Clarion Communications really get eight hundred respondents to an internal email survey for their research, where they knew the result they wanted beforehand, and where Jessica Alba came seventh, but was mysteriously promoted to first after the analysis? Yes, maybe: Clarion is part of WPP, one of the world's largest 'communications services' groups. It does advertising, PR and lobbying, has a turnover of around £6 billion, and employs 100,000 people in a hundred countries.

These corporations run our culture, and they riddle it with bullshit.

Stats, miracle cures and hidden scares

How can we explain the hopelessness of media coverage of science? A lack of expertise is one part of the story, but there are other, more interesting elements. Over half of all the science coverage in a newspaper is concerned with health, because stories of what will kill or cure us are highly motivating, and in this field the pace of research has changed dramatically, as I have already briefly mentioned. This is important background.

Before 1935 doctors were basically useless. We had morphine for pain relief – a drug with superficial charm, at least – and we could do operations fairly cleanly, although with huge doses of anaesthetics, because we hadn't yet sorted out well-targeted muscle-relaxant drugs. Then suddenly, between about 1935 and 1975, science poured out an almost constant stream of miracle cures. If you got TB in the 1920s, you died, pale and emaciated, in the style of a romantic poet. If you got TB in the 1970s, then in all likelihood you would live to a ripe old age. You might have to take rifampicin and isoniazid for months on end, and they're not nice drugs, and the side-effects will make your eyeballs and wee go pink, but if all goes well you will live to see inventions unimaginable in your childhood.

It wasn't just the drugs. Everything we associate with modern medicine happened in that time, and it was a barrage of miracles: kidney dialysis machines allowed people to live on despite losing two vital organs. Transplants brought people back from a death sentence. CT scanners could give three-dimensional images of the inside of a living person. Heart surgery rocketed forward. Almost every drug you've ever heard of was invented. Cardiopulmonary resuscitation (the business with the chest compressions and the electric shocks to bring you back) began in earnest.

Let's not forget polio. The disease paralyses your muscles, and if it affects those of your chest wall, you literally cannot pull air in and out: so you die. Well, reasoned the doctors, polio paralysis often retreats spontaneously. Perhaps, if you could just keep these patients breathing somehow, for weeks on end if necessary, with mechanical ventilation, a bag and a mask, then they might, with time, start to breathe independently once more. They were right. People almost literally came back from the dead, and so intensive care units were born.

Alongside these absolute undeniable miracles, we really were finding those simple, direct, hidden killers that the media still pine for so desperately in their headlines. In 1950 Richard Doll and Austin Bradford Hill published a preliminary 'case-control study' – where you gather cases of people with a particular disease, and find similar people who don't have it, and compare the lifestyle risk factors between the groups – which showed a strong relationship between lung cancer and smoking. The British Doctors Study in 1954 looked at 40,000 doctors – medics are good to study, because they're on the GMC register, so you can find them again easily to see what happened later in their life – and confirmed the finding. Doll and Bradford Hill had been wondering if lung cancer might be related to tarmac, or petrol; but smoking, to everybody's genuine surprise, turned out to cause it in 97 per cent of cases. You will find a massive distraction on the subject in this footnote.*

The golden age – mythical and simplistic though that model may be – ended in the 1970s. But medical research did not grind to a halt. Far from it: your chances of dying as a middle-aged man have probably halved over the past thirty years, but this is not because of any single, dramatic, headline-grabbing break-through. Medical academic research today moves forward through the gradual emergence of small incremental improve-ments, in our understanding of drugs, their dangers and bene-

---

* In some ways, perhaps it shouldn't have been a surprise. The Germans had identified a rise in lung cancer in the 1920s, but suggested – quite reasonably – that it might be related to poison-gas exposure in the Great War. In the 1930s, identifying toxic threats in the environment became an important feature of the Nazi project to build a master race through 'racial hygiene'.

Two researchers, Schairer and Schöniger, published their own case-control study in 1943, demonstrating a relationship between smoking and lung cancer almost a decade before any researchers elsewhere. Their paper wasn't mentioned in the classic Doll and Bradford Hill paper of 1950, and if you check in the Science Citation Index, it was referred to only four times in the 1960s, once in the 1970s, and then not again until 1988, despite providing valuable information. Some might argue that this shows the danger of dismissing sources you dislike. But Nazi scientific and medical research was bound up with the horrors of cold-blooded mass murder, and the strange puritanical ideologies of Nazism. It was almost universally disregarded, and with good reason. Doctors had been active participants in the Nazi project, and joined Hitler's National Socialist Party in greater numbers than any other profession (45 per cent of them were party members, compared with 20 per cent of teachers).

German scientists involved in the smoking project included racial theorists, but also researchers interested in the heritability of frailties created by tobacco, and the question of whether people could be rendered 'degenerate' by their environment. Research on smoking was directed by Karl Astel, who helped to organise the 'euthanasia' operation that murdered 200,000 mentally and physically disabled people, and assisted in the 'final solution of the Jewish question' as head of the Office of Racial Affairs.

fits, best practice in their prescription, the nerdy refinement of obscure surgical techniques, identification of modest risk factors, and their avoidance through public health programmes (like 'five-a-day') which are themselves hard to validate.

This is the major problem for the media when they try to cover medical academic research these days: you cannot crowbar these small incremental steps – which in the aggregate make a sizeable contribution to health – into the pre-existing 'miracle-cure-hidden-scare' template.

I would go further, and argue that science itself works very badly as a news story: it is by its very nature a subject for the 'features' section, because it does not generally move ahead by sudden, epoch-making breakthroughs. It moves ahead by gradually emergent themes and theories, supported by a raft of evidence from a number of different disciplines on a number of different explanatory levels. Yet the media remain obsessed with 'new breakthroughs'.

It's quite understandable that newspapers should feel it's their job to write about new stuff. But if an experimental result is newsworthy, it can often be for the same reasons that mean it is probably wrong: it must be new, and unexpected, it must change what we previously thought; which is to say, it must be a single, lone piece of information which contradicts a large amount of pre-existing experimental evidence.

There has been a lot of excellent work done, much of it by a Greek academic called John Ioannidis, demonstrating how and why a large amount of brand-new research with unexpected results will subsequently turn out to be false. This is clearly important in the application of scientific research to everyday work, for example in medicine, and I suspect most people intuitively understand that: you would be unwise to risk your life on a single piece of unexpected data that went against the grain.

In the aggregate, these 'breakthrough' stories sell the idea that science – and indeed the whole empirical world view – is only about tenuous, new, hotly contested data and spectacular breakthroughs. This reinforces one of the key humanities graduates' parodies of science: as well as being irrelevant boffinry, science is temporary, changeable, constantly revising itself, like a transient fad. Scientific findings, the argument goes, are therefore dismissible.

While this is true at the bleeding edges of various research fields, it's worth bearing in mind that Archimedes has been right about why things float for a couple of millennia. He also understood why levers work, and Newtonian physics will probably be right about the behaviour of snooker balls forever.* But somehow this impression about the changeability of science has bled through to the core claims. Anything can be rubbished.

But that is all close to hand-waving. We should now look at how the media cover science, unpick the real meanings behind the phrase 'research has shown', and, most importantly of all, examine the ways in which the media repeatedly and routinely misrepresent and misunderstand statistics.

'Research has shown ...'

The biggest problem with science stories is that they routinely contain no scientific evidence at all. Why? Because papers think you won't understand the 'science bit', so all stories involving science must be dumbed down, in a desperate bid to seduce and engage the ignorant, who are not interested in science anyway (perhaps because journalists think it is good for you, and so should be democratised).

In some respects these are admirable impulses, but there are certain inconsistencies I can't help noticing. Nobody dumbs down the finance pages. I can barely understand most of the

---

* I cheerfully admit to borrowing these examples from fabulous Professor Lewis Wolpert.

sports section. In the literature pull-out, there are five-page-long essays which I find completely impenetrable, where the more Russian novelists you can rope in the cleverer everybody thinks you are. I do not complain about this: I envy it.

If you are simply presented with the conclusions of a piece of research, without being told what was measured, how, and what was found – the evidence – then you are simply taking the researchers' conclusions at face value, and being given no insight into the process. The problems with this are best explained by a simple example.

Compare the two sentences 'Research has shown that black children in America tend to perform less well in IQ tests than white children' and 'Research has shown that black people are less intelligent than white people.' The first tells you about what the research found: it is the evidence. The second tells you the hypothesis, somebody's interpretation of the evidence: somebody who, you will agree, doesn't know much about the relationship between IQ tests and intelligence.

With science, as we have seen repeatedly, the devil is in the detail, and in a research paper there is a very clear format: you have the methods and results section, the meat, where you describe what was done, and what was measured; and then you have the conclusions section, quite separately, where you give your impressions, and mesh your findings with those of others to decide whether they are compatible with each other, and with a given theory. Often you cannot trust researchers to come up with a satisfactory conclusion on their results – they might be really excited about one theory – and you need to check their actual experiments to form your own view. This requires that news reports are about published research which can, at least, be read somewhere. It is also the reason why publication in full – and review by anyone in the world who wants to read your paper – is more important than 'peer review', the process whereby academic journal articles are given the once-over by a few academics working in the field, checking for gross errors and the like.

In the realm of their favourite scares, there is a conspicuous over-reliance by newspapers on scientific research that has not been published at all. This is true of almost all of the more recent headline stories on new MMR research, for example. One regularly quoted source, Dr Arthur Krigsman, has been making widely reported claims for new scientific evidence on MMR since 2002, and has yet to publish his work in an academic journal to this day, six years later. Similarly, the unpublished 'GM potato' claims of Dr Arpad Pusztai that genetically modified potatoes caused cancer in rats resulted in 'Frankenstein food' headlines for a whole year before the research was finally published, and could be read and meaningfully assessed. Contrary to the media speculation, his work did not support the hypothesis that GM is injurious to health (this doesn't mean it's necessarily a good thing – as we will see later).

Once you become aware of the difference between the evidence and the hypothesis, you start to notice how very rarely you get to find out what any research has really shown when journalists say 'research has shown'.

Sometimes it's clear that the journalists themselves simply don't understand the unsubtle difference between the evidence and the hypothesis. *The Times*, for example, covered an experiment which showed that having younger siblings was associated with a lower incidence of multiple sclerosis. MS is caused by the immune system turning on the body. 'This is more likely to happen if a child at a key stage of development is not exposed to infections from younger siblings, says the study.' That's what *The Times* said.

But it's wrong. That's the 'hygiene hypothesis', that's the theory, the framework into which the evidence might fit, but it's not what the study showed: the study just found that having younger siblings seemed to be somewhat protective against MS.

It didn't say what the mechanism was, it couldn't say why there was a relationship, such as whether it happened through greater exposure to infections. It was just an observation. *The Times* confused the evidence with hypothesis, and I am very glad to have got that little gripe out of my system.

How do the media work around their inability to deliver scientific evidence? Often they use authority figures, the very antithesis of what science is about, as if they were priests, or politicians, or parent figures. 'Scientists today said ... Scientists revealed ... Scientists warned'. If they want balance, you'll get two scientists disagreeing, although with no explanation of why (an approach which can be seen at its most dangerous in the myth that scientists were 'divided' over the safety of MMR). One scientist will 'reveal' something, and then another will 'challenge' it. A bit like Jedi knights.

There is a danger with authority-figure coverage, in the absence of real evidence, because it leaves the field wide open for questionable authority figures to waltz in. Gillian McKeith, Andrew Wakefield and the rest can all get a whole lot further in an environment where their authority is taken as read, because their reasoning and evidence are rarely publicly examined.

Worse than that, where there is controversy about what the evidence shows, it reduces the discussion to a slanging match, because a claim such as 'MMR causes autism' (or not), is only critiqued in terms of the *character* of the person who is making the statement, rather than the evidence they are able to present. There is no need for this, as we shall see, because people are not stupid, and the evidence is often pretty easy to understand.

It also reinforces the humanities graduate journalists' parody of science, for which we now have all the ingredients: science is about groundless, changeable, didactic truth statements from arbitrary unelected authority figures. When they start to write about serious issues like MMR, you can see that this is what people in the media really think science is about. The next stop on our journey is inevitably going to be statistics, because this is one area that causes unique problems for the media. But first, we need to go on a brief diversion.

12

Why Clever People Believe Stupid Things

The real purpose of the scientific method is to make sure nature hasn't misled you into thinking you know something you actually don't know.

Robert Pirsig, Zen and the Art of Motorcycle Maintenance

Why do we have statistics, why do we measure things, and why do we count? If the scientific method has any authority – or as I prefer to think of it, 'value' – it is because it represents a systematic approach; but this is valuable only because the alternatives can be misleading. When we reason informally – call it intuition, if you like – we use rules of thumb which simplify problems for the sake of efficiency. Many of these shortcuts have been well characterised in a field called 'heuristics', and they are efficient ways of knowing in many circumstances.

This convenience comes at a cost – false beliefs – because there are systematic vulnerabilities in these truth-checking strategies which can be exploited. This is not dissimilar to the way that paintings can exploit shortcuts in our perceptual system: as objects become more distant, they appear smaller, and 'perspective' can trick us into seeing three dimensions where there are only two, by taking advantage of this strategy used by our depth-checking apparatus. When our cognitive system – our truth-checking apparatus – is fooled, then, much like seeing depth in a flat painting, we come to erroneous conclusions about abstract things. We might misidentify normal fluctuations as meaningful patterns, for example, or ascribe causality where in fact there is none.

These are cognitive illusions, a parallel to optical illusions. They can be just as mind-boggling, and they cut to the core of why we do science, rather than basing our beliefs on intuition informed by a 'gist' of a subject acquired through popular media: because the world does not provide you with neatly tabulated data on interventions and outcomes. Instead it gives you random, piecemeal data in dribs and drabs over time, and trying to construct a broad understanding of the world from a memory of your own experiences would be like looking at the ceiling of the Sistine Chapel through a long, thin cardboard tube: you can try to remember the individual portions you've spotted here and there, but without a system and a model, you're never going to appreciate the whole picture.

Let's begin.

Randomness

As human beings, we have an innate ability to make something out of nothing. We see shapes in the clouds, and a man in the moon; gamblers are convinced that they have 'runs of luck'; we take a perfectly cheerful heavy-metal record, play it backwards, and hear hidden messages about Satan. Our ability to spot patterns is what allows us to make sense of the world; but sometimes, in our eagerness, we are oversensitive, trigger-happy, and mistakenly spot patterns where none exist.

In science, if you want to study a phenomenon, it is sometimes useful to reduce it to its simplest and most controlled form. There is a prevalent belief among sporting types that sportsmen, like gamblers (except more plausibly), have 'runs of luck'. People ascribe this to confidence, 'getting your eye in', 'warming up', or more, and while it might exist in some games, statisticians have looked in various places where people have claimed it to exist and found no relationship between, say, hitting a home run in one inning, then hitting a home run in the next.

Because the 'winning streak' is such a prevalent belief, it is an excellent model for looking at how we perceive random sequences of events. This was used by an American social psychologist called Thomas Gilovich in a classic experiment. He took basketball fans and showed them a random sequence of X's and O's, explaining that they represented a player's hits and misses, and then asked them if they thought the sequences demonstrated 'streak shooting'.

Here is a random sequence of figures from that experiment. You might think of it as being generated by a series of coin tosses.

OXXXOXXXOXXOOOXOOXXOO

The subjects in the experiment were convinced that this sequence exemplified 'streak shooting' or 'runs of luck', and it's easy to see why, if you look again: six of the first eight shots were hits. No, wait: eight of the first eleven shots were hits. No way is that random …

What this ingenious experiment shows is how bad we are at correctly identifying random sequences. We are wrong about what they should look like: we expect too much alternation, so truly random sequences seem somehow too lumpy and ordered. Our intuitions about the most basic observation of all – distinguishing a pattern from mere random background noise – are deeply flawed.

This is our first lesson in the importance of using statistics instead of intuition. It's also an excellent demonstration of how strong the parallels are between these cognitive illusions and the perceptual illusions with which we are more familiar. You can stare at a visual illusion all you like, talk or think about it, but it will still look 'wrong'. Similarly, you can look at that random sequence above as hard as you like: it will still look lumpy and ordered, in defiance of what you now know.

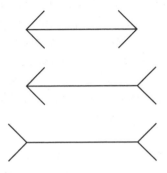

Regression to the mean

We have already looked at regression to the mean in our section on homeopathy: it is the phenomenon whereby, when things are at their extremes, they are likely to settle back down to the middle, or 'regress to the mean'.

We saw this with reference to the *Sports Illustrated* jinx (and Bruce Forsyth's *Play Your Cards Right*), but also applied it to the matter in hand, the question of people getting better: we discussed how people will do something when their back pain is at its worst – visit a homeopath, perhaps – and how although it was going to get better anyway (because when things are at their worst they generally do), they ascribe their improvement to the treatment.

There are two discrete things happening when we fall prey to this failure of intuition. Firstly, we have failed to correctly spot the pattern of regression to the mean. Secondly, crucially, we have then decided that something must have *caused* this illusory pattern: specifically, a homeopathic remedy, for example. Simple regression is confused with causation, and this is perhaps quite natural for animals like humans, whose success in the world depends on our being able to spot causal relationships rapidly and intuitively: we are inherently oversensitive to them.

To an extent, when we discussed the subject earlier I relied on your good will, and on the likelihood that from your own experience you could agree that this explanation made sense. But it has been demonstrated in another ingeniously pared-down experiment, where all the variables were controlled, but people still saw a pattern, and causality, where there was none.

The subjects in the experiment played the role of a teacher trying to make a child arrive punctually at school for 8.30 a.m. They sat at a computer on which it appeared that each day, for fifteen consecutive days, the supposed child would arrive some time between 8.20 and 8.40; but unbeknownst to the subjects, the arrival times were entirely random, and predetermined before the experiment began. Nonetheless, the subjects were all allowed to use punishments for lateness, and rewards for punctuality, in whatever permutation they wished. When they were asked at the end to rate their strategy, 70 per cent concluded that reprimand was more effective than reward in producing punctuality from the child.

These subjects were convinced that their intervention had an effect on the punctuality of the child, despite the child's arrival time being entirely random, and exemplifying nothing more than 'regression to the mean'. By the same token, when homeopathy has been shown to elicit no more improvement than placebo, people are still convinced that it has a beneficial effect on their health.

To recap:

1. We see patterns where there is only random noise.
2. We see causal relationships where there are none.

These are two very good reasons to measure things formally. It's bad news for intuition already. Can it get much worse?

The bias towards positive evidence

It is the peculiar and perpetual error of the human understanding to be more moved and excited by affirmatives than negatives.

Francis Bacon

It gets worse. It seems we have an innate tendency to seek out and overvalue evidence that confirms a given hypothesis. To try to remove this phenomenon from the controversial arena of CAM – or the MMR scare, which is where this is headed – we are lucky to have more pared-down experiments which illustrate the general point.

Imagine a table with four cards on it, marked 'A', 'B', '2' and '3'. Each card has a letter on one side, and a number on the other. Your task is to determine whether all cards with a vowel on one side have an even number on the other. Which two cards would you turn over? Everybody chooses the 'A' card, obviously, but like many people – unless you really forced yourself to think hard about it – you would probably choose to turn over the '2' card as well. That's because these are the cards which would produce information *consistent* with the hypothesis you are supposed to be testing. But in fact, the cards you need to flip are the 'A' and the '3', because finding a vowel on the back of the '2' would tell you nothing about 'all cards', it would just confirm 'some cards', whereas finding a vowel on the back of '3' would comprehensively disprove your hypothesis. This modest brain-teaser demonstrates our tendency, in our unchecked intuitive reasoning style, to seek out information that confirms a hypothesis: and it demonstrates the phenomenon in a value-neutral situation.

This same bias in seeking out confirmatory information has been demonstrated in more sophisticated social psychology experiments. When trying to determine if someone is an 'extrovert', for example, many subjects will ask questions for which a positive answer would confirm the hypothesis ('Do you like going to parties?') rather than refute it.

We show a similar bias when we interrogate information from our own memory. In one experiment, subjects read a vignette about a woman who exemplified various introverted and extroverted behaviours, and were then divided into two groups. One group was asked to consider her suitability for a job as a librarian, while the other was asked to consider her suitability for a job as an estate agent. Both groups were asked to come up with examples of both her extroversion and her introversion. The group considering her for the librarian job recalled more examples of introverted behaviour, while the group considering her for a job selling real estate cited more examples of extroverted behaviour.

This tendency is dangerous, because if you only ask questions that confirm your hypothesis, you will be more likely to elicit information that confirms it, giving a spurious sense of confirmation. It also means – thinking more broadly – that the people who pose the questions already have a head start in popular discourse.

So we can add to our running list of cognitive illusions, biases and failings of intuition:

3. We overvalue confirmatory information for any given hypothesis.
4. We seek out confirmatory information for any given hypothesis.

Biased by our prior beliefs

[I] followed a golden rule, whenever a new observation or thought came across me, which was opposed to my general results, to make a memorandum of it without fail and at once; for I had found by experience that such facts and thoughts were far more apt to escape from the memory than favourable ones.

Charles Darwin

This is the reasoning flaw that everybody does know about, and even if it's the least interesting cognitive illusion – because it's an obvious one – it has been demonstrated in experiments which are so close to the bone that you may find them, as I do, quite unnerving.

The classic demonstration of people being biased by their prior beliefs comes from a study looking at beliefs about the death penalty. A large number of proponents and opponents of state executions were collected. They were all shown two pieces of evidence on the deterrent effect of capital punishment: one supporting a deterrent effect, the other providing evidence against it.

The evidence they were shown was as follows:
- A comparison of murder rates in one US state before the death penalty was brought in, and after.
- A comparison of murder rates in different states, some with, and some without, the death penalty.

But there was a very clever twist. The proponents and opponents of capital punishment were each further divided into two smaller groups. So, overall, half of the proponents and opponents of capital punishment had their opinion reinforced by before/after data, but challenged by state/state data, and vice versa.

Asked about the evidence, the subjects confidently uncovered flaws in the methods of the research that went against their pre-existing view, but downplayed the flaws in the research that supported their view. Half the proponents of capital punishment, for example, picked holes in the idea of state/state comparison data, on methodological grounds, because that was the data that went against their view, while they were happy with the before/after data; but the other half of the proponents of capital punishment rubbished the before/after data, because in their case they had been exposed to before/after data which challenged their view, and state/state data which supported it.

Put simply, the subjects' faith in research data was not predicated on an objective appraisal of the research methodology, but on whether the results validated their pre-existing views. This phenomenon reaches its pinnacle in alternative therapists – or scaremongers – who unquestioningly champion anecdotal data, whilst meticulously examining every large, carefully conducted study on the same subject for any small chink that would permit them to dismiss it entirely.

This, once again, is why it is so important that we have clear strategies available to us to appraise evidence, regardless of its conclusions, and this is the major strength of science. In a systematic review of the scientific literature, investigators will sometimes mark the quality of the 'methods' section of a study blindly – that is, without looking at the 'results' section – so that it cannot bias their appraisal. Similarly, in medical research there is a hierarchy of evidence: a well performed trial is more significant than survey data in most contexts, and so on.

So we can add to our list of new insights about the flaws in intuition:

5. Our assessment of the quality of new evidence is biased by our previous beliefs.

Availability

We spend our lives spotting patterns, and picking out the exceptional and interesting things. You don't waste cognitive effort, every time you walk into your house, noticing and analysing all the many features in the visually dense environment of your kitchen. You do notice the broken window and the missing telly.

When information is made more 'available', as psychologists call it, it becomes disproportionately prominent. There are a number of ways this can happen, and you can pick up a picture of them from a few famous psychology experiments into the phenomenon.

In one, subjects were read a list of male and female names, in equal number, and then asked at the end whether there were more men or women in the list: when the men in the list had names like Ronald Reagan, but the women were unheard of, people tended to answer that there were more men than women; and vice versa.

Our attention is always drawn to the exceptional and the interesting, and if you have something to sell, it makes sense to guide people's attention to the features you most want them to notice. When fruit machines pay up, they make a theatrical 'kerchunk-kerchunk' sound with every coin they spit out, so that everybody in the pub can hear it; but when you lose, they don't draw attention to themselves. Lottery companies, similarly, do their absolute best to get their winners prominently into the media; but it goes without saying that you, as a lottery loser, have never had your outcome paraded for the TV cameras.

The anecdotal success stories about CAM – and the tragic anecdotes about the MMR vaccine – are disproportionately misleading, not just because the statistical context is missing, but because of their 'high availability': they are dramatic, associated with strong emotion, and amenable to strong visual imagery. They are concrete and memorable, rather than abstract. No matter what you do with statistics about risk or recovery, your numbers will always have inherently low psychological availability, unlike miracle cures, scare stories, and distressed parents.

It's because of 'availability', and our vulnerability to drama, that people are more afraid of sharks at the beach, or of fairground rides on the pier, than they are of flying to Florida, or driving to the coast. This phenomenon is even demonstrated in patterns of smoking cessation amongst doctors: you'd imagine, since they are rational actors, that all doctors would simultaneously have seen sense and stopped smoking once they'd read the studies showing the phenomenally compelling relationship between cigarettes and lung cancer. These are men of applied science, after all, who are able, every day, to translate cold statistics into meaningful information and beating human hearts.

But in fact, from the start, doctors working in specialities like chest medicine and oncology – where they witnessed patients dying of lung cancer with their own eyes – were proportionately more likely to give up cigarettes than their colleagues in other specialities. Being shielded from the emotional immediacy and drama of consequences matters.

Social influences

Last in our whistle-stop tour of irrationality comes our most self-evident flaw. It feels almost too obvious to mention, but our values are socially reinforced by conformity and by the company we keep. We are selectively exposed to information that revalidates our beliefs, partly because we expose ourselves to *situations* where those beliefs are apparently confirmed; partly because we ask questions that will – by their very nature, for the reasons described above – give validating answers; and partly because we selectively expose ourselves to *people* who validate our beliefs.

It's easy to forget the phenomenal impact of conformity. You doubtless think of yourself as a fairly independent-minded person, and you know what you think. I would suggest that the same beliefs were held by the subjects of Asch's experiments into social conformity. These subjects were placed near one end of a line of actors who presented themselves as fellow experimental subjects, but were actually in cahoots with the experimenters. Cards were held up with one line marked on them, and then another card was held up with three lines of different lengths: six inches, eight inches, ten inches.

Everyone called out in turn which line on the second card was the same length as the line on the first. For six of the eighteen pairs of cards the accomplices gave the correct answer; but for the other twelve they called out the wrong answer. In all but a quarter of the cases, the experimental subjects went along with the incorrect answer from the crowd of accomplices on one or more occasions, defying the clear evidence of their own senses.

That's an extreme example of conformity, but the phenomenon is all around us. 'Communal reinforcement' is the process by which a claim becomes a strong belief, through repeated assertion by members of a community. The process is independent of whether the claim has been properly researched, or is supported by empirical data significant enough to warrant belief by reasonable people.

Communal reinforcement goes a long way towards explaining how religious beliefs can be passed on in communities from generation to generation. It also explains how testimonials within communities of therapists, psychologists, celebrities, theologians, politicians, talk-show hosts, and so on, can supplant and become more powerful than scientific evidence.

When people learn no tools of judgement and merely follow their hopes, the seeds of political manipulation are sown.

Stephen Jay Gould

There are many other well-researched areas of bias. We have a disproportionately high opinion of ourselves, which is nice. A large majority of the public think they are more fair-minded, less prejudiced, more intelligent and more skilled at driving than the average person, when of course only half of us can be better than the median.* Most of us exhibit something called 'attributional bias': we believe our successes are due to our own internal faculties, and our failures are due to external factors; whereas for others, we believe their successes are due to luck, and their failures to their own flaws. We can't all be right.

Lastly, we use context and expectation to bias our appreciation of a situation – because, in fact, that's the only way we can think. Artificial intelligence research has drawn a blank so far largely because of something called the 'frame problem': you can tell a computer how to process information, and give it all the information in the world, but as soon as you give it a real-world problem – a sentence to understand and respond to, for example – computers perform much worse than we might expect, because they don't know what information is relevant to the problem. This is something humans are very good at – filtering irrelevant information – but that skill comes at a cost of ascribing disproportionate bias to some contextual data.

We tend to assume, for example, that positive characteristics cluster: people who are attractive must also be good; people who seem kind might also be intelligent and well-informed. Even this has been demonstrated experimentally: identical essays in neat handwriting score higher than messy ones; and the behaviour of sporting teams which wear black is rated as more aggressive and unfair than teams which wear white.

---

* I'd be genuinely intrigued to know how long it takes to find someone who can tell you the difference between 'median', 'mean' and 'mode', from where you are sitting right now.

And no matter how hard you try, sometimes things just are very counterintuitive, especially in science. Imagine there are twenty-three people in a room. What is the chance that two of them celebrate their birthday on the same date? One in two.*

When it comes to thinking about the world around you, you have a range of tools available. Intuitions are valuable for all kinds of things, especially in the social domain: deciding if your girlfriend is cheating on you, perhaps, or whether a business partner is trustworthy. But for mathematical issues, or assessing causal relationships, intuitions are often completely wrong, because they rely on shortcuts which have arisen as handy ways to solve complex cognitive problems rapidly, but at a cost of inaccuracies, misfires and oversensitivity.

It's not safe to let our intuitions and prejudices run unchecked and unexamined: it's in our interest to challenge these flaws in intuitive reasoning wherever we can, and the methods of science and statistics grew up specifically in opposition to these flaws. Their thoughtful application is our best weapon against these pitfalls, and the challenge, perhaps, is to work out which tools to use where. Because trying to be 'scientific' about your relationship with your partner is as stupid as following your intuitions about causality.

Now let's see how journalists deal with stats.

---

* If it helps to make this feel a bit more plausible, bear in mind that you only need *any* two dates to coincide. With forty-seven people, the probability increases to 0.95: that's nineteen times out of twenty! (Fifty-seven people and it's 0.99; seventy people and it's 0.999.) This is beyond your intuition: at first glance, it makes no sense at all.

13

Bad Stats

Now that you appreciate the value of statistics – the benefits and risks of intuition – we can look at how these numbers and calculations are repeatedly misused and misunderstood. Our first examples will come from the world of journalism, but the true horror is that journalists are not the only ones to make basic errors of reasoning.

Numbers, as we will see, can ruin lives.

The biggest statistic
Newspapers like big numbers and eye-catching headlines. They need miracle cures and hidden scares, and small percentage shifts in risk will never be enough for them to sell readers to advertisers (because that is the business model). To this end they pick the single most melodramatic and misleading way of describing any statistical increase in risk, which is called the 'relative risk increase'.

Let's say the risk of having a heart attack in your fifties is 50 per cent higher if you have high cholesterol. That sounds pretty bad. Let's say the extra risk of having a heart attack if you have high cholesterol is only 2 per cent. That sounds OK to me. But they're the same (hypothetical) figures. Let's try this. Out of a hundred men in their fifties with normal cholesterol, four will be expected to have a heart attack; whereas out of a hundred men with high cholesterol, six will be expected to have a heart attack. That's two extra heart attacks per hundred. Those are called 'natural frequencies'.

Natural frequencies are readily understandable, because instead of using probabilities, or percentages, or anything even slightly technical or difficult, they use concrete numbers, just like the ones you use every day to check if you've lost a kid on a coach trip, or got the right change in a shop. Lots of people have argued that we evolved to reason and do maths with concrete numbers like these, and not with probabilities, so we find them more intuitive. Simple numbers are simple.

The other methods of describing the increase have names too. From our example above, with high cholesterol, you could have a 50 per cent increase in risk (the 'relative risk increase'); or a 2 per cent increase in risk (the 'absolute risk increase'); or, let me ram it home, the easy one, the informative one, an extra two heart attacks for every hundred men, the natural frequency.

As well as being the most comprehensible option, natural frequencies also contain more information than the journalists' 'relative risk increase'. Recently, for example, we were told that red meat causes bowel cancer, and ibuprofen increases the risk of heart attacks: but if you followed the news reports, you would be no wiser. Try this, on bowel cancer, from the *Today* programme on Radio 4: 'A bigger risk meaning what, Professor Bingham?' 'A third higher risk.' 'That sounds an awful lot, a third higher risk; what are we talking about in terms of numbers here?' 'A difference ... of around about twenty people per year.' 'So it's still a small number?' 'Umm ... per 10,000 ...'

These things are hard to communicate if you step outside of the simplest format. Professor Sheila Bingham is Director of the MRC Centre for Nutrition in Cancer Epidemiology Prevention and Survival at the University of Cambridge, and deals with these numbers for a living, but in this (entirely forgivable) fumbling on a live radio show she is not alone: there are studies of doctors, and commissioning committees for local health authorities, and members of the legal profession, which show that people who interpret and manage risk for a living often have huge difficulties expressing what they mean on the spot. They are also much more likely to make the right decision when information about risk is presented as natural frequencies, rather than as probabilities or percentages.

For painkillers and heart attacks, another front-page story, the desperate urge to choose the biggest possible number led to the figures being completely inaccurate, in many newspapers. The reports were based on a study that had observed participants over four years, and the results suggested, using natural frequencies, that you would expect one extra heart attack for every 1,005 people taking ibuprofen. Or as the *Daily Mail*, in an article titled 'How Pills for Your Headache Could Kill', reported: 'British research revealed that patients taking ibuprofen to treat arthritis face a 24 per cent increased risk of suffering a heart attack.' Feel the fear.

Almost everyone reported the relative risk increases: diclofenac increases the risk of heart attack by 55 per cent, ibuprofen by 24 per cent. Only the *Daily Telegraph* and the *Evening Standard* reported the natural frequencies: one extra heart attack in 1,005 people on ibuprofen. The *Mirror*, meanwhile, tried and failed, reporting that one in 1,005 people on ibuprofen 'will suffer heart failure over the following year'. No. It's heart attack, not heart failure, and it's one *extra* person in 1,005, on top of the heart attacks you'd get anyway. Several other papers repeated the same mistake.

Often it's the fault of the press releases, and academics can themselves be as guilty as the rest when it comes to overdramatising their research (there are excellent best-practice guidelines from the Royal Society on communicating research, if you are interested). But if anyone in a position of power is reading this, here is the information I would like from a newspaper, to help me make decisions about my health, when reporting on a risk: I want to know who you're talking about (e.g. men in their fifties); I want to know what the baseline risk is (e.g. four men out of a hundred will have a heart attack over ten years); and I want to know what the increase in risk is, as a natural frequency (two extra men out of that hundred will have a heart attack over ten years). I also want to know exactly what's causing that increase in risk – an occasional headache pill or a daily tub full of pain-relieving medication for arthritis. Then I will consider reading your newspapers again, instead of blogs which are written by people who understand research, and which link reliably back to the original academic paper, so that I can double-check their précis when I wish.

Over a hundred years ago, H.G. Wells said that statistical thinking would one day be as important as the ability to read and write in a modern technological society. I disagree; probabilistic reasoning is difficult for everyone, but everyone understands normal numbers. This is why 'natural frequencies' are the only sensible way to communicate risk.

Choosing your figures

Sometimes the mispresentation of figures goes so far beyond reality that you can only assume mendacity. Often these situations seem to involve morality: drugs, abortion and the rest. With very careful selection of numbers, in what some might consider to be a cynical and immoral manipulation of the facts for personal gain, you can sometimes make figures say anything you want.

The *Independent* was in favour of legalising cannabis for many years, but in March 2007 it decided to change its stance. One option would have been simply to explain this as a change of heart, or a reconsideration of the moral issues. Instead it was decorated with science – as cowardly zealots have done from eugenics through to prohibition – and justified with a fictitious change in the facts. 'Cannabis – An Apology' was the headline for their front-page splash.

> In 1997, this newspaper launched a campaign to decriminalise the drug. If only we had known then what we can reveal today … Record numbers of teenagers are requiring drug treatment as a result of smoking skunk, the highly potent cannabis strain that is 25 times stronger than resin sold a decade ago.

Twice in this story we are told that cannabis is twenty-five times stronger than it was a decade ago. For the paper's former editor Rosie Boycott, in her melodramatic recantation, skunk was 'thirty times stronger'. In one inside feature the strength issue was briefly downgraded to a 'can be'. The paper even referenced its figures: 'The Forensic Science Service says that in the early Nineties cannabis would contain around 1 per cent tetrahydrocannabidinol (THC), the mind-altering compound, but can now have up to 25 per cent.'

This is all sheer fantasy.

I've got the Forensic Science Service data right here in front of me, and the earlier data from the Laboratory of the Government Chemist, the United Nations Drug Control Program, and the European Union's Monitoring Centre for Drugs and Drug Addiction. I'm going to share it with you, because I happen to think that people are very well able to make their own minds up about important social and moral issues when given the facts.

The data from the Laboratory of the Government Chemist goes from 1975 to 1989. Cannabis resin pootles around between 6 per cent and 10 per cent THC, herbal between 4 per cent and 6 per cent. There is no clear trend.

The Forensic Science Service data then takes over to produce the more modern figures, showing not much change in resin,

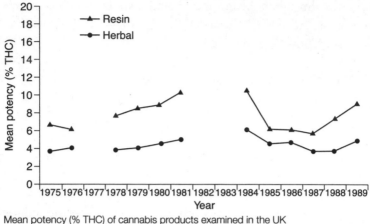

Mean potency (% THC) of cannabis products examined in the UK
(Laboratory of the Government Chemist, 1975–89)

and domestically produced indoor herbal cannabis doubling in potency from 6 per cent to around 12 or 14 per cent. (2003–05 data in table under references).

The rising trend of cannabis potency is gradual, fairly unspectacular, and driven largely by the increased availability of domestic, intensively grown indoor herbal cannabis.

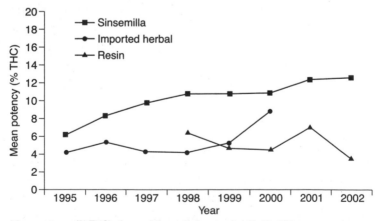

Mean potency (% THC) of cannabis products examined in the UK
(Forensic Science Service, 1995–2002)

Bad Stats

Year	Sinsemilla %	Resin %	'Traditional' imported herbal %
1995	5.8	No data	3.9
1996	8.0	No data	5.0
1997	9.4	No data	4.0
1998	10.5	6.1	3.9
1999	10.6	4.4	5.0
2000	12.2	4.2	8.5
2001	12.3	6.7	No data
2002	12.3	3.2	No data
2003	12.0	4.6	No data
2004	12.7	1.6	No data
2005	14.2	6.6	No data

Mean THC content of cannabis products seized in the UK
(Forensic Science Service, 1995–2002)

'Twenty-five times stronger', remember. Repeatedly, and on the front page.

If you were in the mood to quibble with the *Independent*'s moral and political reasoning, as well as its evident and shameless venality, you could argue that intensive indoor cultivation of a plant which grows perfectly well outdoors is the cannabis industry's reaction to the product's illegality itself. It is dangerous to import cannabis in large amounts. It is dangerous to be caught growing a field of it. So it makes more sense to grow it intensively indoors, using expensive real estate, but producing a more concentrated drug. More concentrated drugs products are, after all, a natural consequence of illegality. You can't buy coca leaves in Peckham, although you can buy crack.

There is, of course, exceptionally strong cannabis to be found in some parts of the British market today, but then there always has been. To get its scare figure, the *Independent* can only have compared the *worst* cannabis from the past with the *best* cannabis of today. It's an absurd thing to do, and moreover you could have cooked the books in exactly the same way thirty years ago if you'd wanted: the figures for individual samples are available, and in 1975 the weakest herbal cannabis analysed was 0.2 per cent THC, while in 1978 the strongest herbal cannabis was 12 per cent. By these figures, in just three years herbal cannabis became 'sixty times stronger'.

And this scare isn't even new. In the mid-1980s, during Ronald Reagan's 'war on drugs' and Zammo's 'Just say no' campaign on *Grange Hill*, American campaigners were claiming that cannabis was fourteen times stronger than in 1970. Which sets you thinking. If it was fourteen times stronger in 1986 than in 1970, and it's twenty-five times stronger today than at the beginning of the 1990s, does that mean it's now 350 times stronger than in 1970?

That's not even a crystal in a plant pot. It's impossible. It would require more THC to be present in the plant than the total volume of space taken up by the plant itself. It would require matter to be condensed into super-dense quark-gluon-plasma cannabis. For God's sake don't tell the *Independent* such a thing is possible.

Cocaine floods the playground

We are now ready to move on to some more interesting statistical issues, with another story from an emotive area, an article in *The Times* in March 2006 headed: 'Cocaine Floods the Playground'. 'Use of the addictive drug by children doubles in a year,' said the subheading. Was this true?

If you read the press release for the government survey on which the story is based, it reports 'almost no change in patterns of drug use, drinking or smoking since 2000'. But this was a government press release, and journalists are paid to investigate: perhaps the press release was hiding something, to cover up for government failures. The *Telegraph* also ran the 'cocaine use doubles' story, and so did the *Mirror*. Did the journalists find the news themselves, buried in the report?

You can download the full document online. It's a survey of 9,000 children, aged eleven to fifteen, in 305 schools. The three-page summary said, again, that there was no change in prevalence of drug use. If you look at the full report you will find the raw data tables: when asked whether they had used cocaine in the past year, 1 per cent said yes in 2004, and 2 per cent said yes in 2005.

So the newspapers were right: it doubled? No. Almost all the figures given were 1 per cent or 2 per cent. They'd all been rounded off. Civil servants are very helpful when you ring them up. The actual figures were 1.4 per cent for 2004, and 1.9 per cent for 2005, not 1 per cent and 2 per cent. So cocaine use hadn't doubled at all. But people were still eager to defend this story: cocaine use, after all, had increased, yes?

No. What we now have is a relative risk increase of 35.7 per cent, or an absolute risk increase of 0.5 per cent. Using the real numbers, out of 9,000 kids we have about forty-five more saying 'Yes' to the question 'Did you take cocaine in the past year?'

Presented with a small increase like this, you have to think: is it statistically significant? I did the maths, and the answer is yes, it is, in that you get a p-value of less than 0.05. What does 'statistically significant' mean? It's just a way of expressing the likelihood that the result you got was attributable merely to chance. Sometimes you might throw 'heads' five times in a row, with a completely normal coin, especially if you kept tossing it for long enough. Imagine a jar of 980 blue marbles, and twenty red ones, all mixed up: every now and then – albeit rarely – picking blindfolded, you might pull out three red ones in a row, just by chance. The standard cut-off point for statistical significance is a p-value of 0.05, which is just another way of saying, 'If I did this experiment a hundred times, I'd expect a spurious positive result on five occasions, just by chance.'

To go back to our concrete example of the kids in the playground, let's imagine that there was definitely no difference in cocaine use, but you conducted the same survey a hundred times: you might get a difference like the one we have seen here, just by chance, just because you randomly happened to pick up more of the kids who had taken cocaine this time around. But you would expect this to happen less than five times out of your hundred surveys.

So we have a risk increase of 35.7 per cent, which seems at face value to be statistically significant; but it is an isolated figure. To 'data mine', taking it out of its real-world context, and saying it is significant, is misleading. The statistical test for significance assumes that every data point is independent, but here the data is 'clustered', as statisticians say. They are not data points, they are real children, in 305 schools. They hang out together, they copy each other, they buy drugs from each other, there are crazes, epidemics, group interactions.

The increase of forty-five kids taking cocaine could have been a massive epidemic of cocaine use in one school, or a few groups of a dozen kids in a few different schools, or mini-epidemics in a handful of schools. Or forty-five kids independently sourcing and consuming cocaine alone without their friends, which seems pretty unlikely to me.

This immediately makes our increase less statistically significant. The small increase of 0.5 per cent was only significant because it came from a large sample of 9,000 data points – like 9,000 tosses of a coin – and the one thing almost everyone knows about studies like this is that a bigger sample size means the results are probably more significant. But if they're not independent data points, then you have to treat it, in some respects, like a smaller sample, so the results become less significant. As statisticians would say, you must 'correct for clustering'. This is done with clever maths which makes everyone's head hurt. All you need to know is that the reasons why you must 'correct for clustering' are transparent, obvious and easy, as we have just seen (in fact, as with many implements, knowing when to use a statistical tool is a different and equally important skill to understanding how it is built). When you correct for clustering, you greatly reduce the significance of the results. Will our increase in cocaine use, already down from 'doubled' to '35.7 per cent', even survive?

No. Because there is a final problem with this data: there is so much of it to choose from. There are dozens of data points in the report: on solvents, cigarettes, ketamine, cannabis, and so on. It is standard practice in research that we only accept a finding as significant if it has a p-value of 0.05 or less. But as we said, a p-value of 0.05 means that for every hundred comparisons you do, five will be positive by chance alone. From this report you could have done dozens of comparisons, and some of them would indeed have shown increases in usage – but by chance alone, and the cocaine figure could be one of those. If you roll a pair of dice often enough, you will get a double six three times in a row on many occasions. This is why statisticians do a 'correction for multiple comparisons', a correction for 'rolling the dice' lots of times. This, like correcting for clustering, is particularly brutal on the data, and often reduces the significance of findings dramatically.

Data dredging is a dangerous profession. You could – at face value, knowing nothing about how stats works – have said that this government report showed a significant increase of 35.7 per cent in cocaine use. But the stats nerds who compiled it knew about clustering, and Bonferroni's correction for multiple comparisons. They are not stupid; they do stats for a living.

That, presumably, is why they said quite clearly in their summary, in their press release and in the full report that there was no change from 2004 to 2005. But the journalists did not want to believe this: they tried to re-interpret the data for themselves, they looked under the bonnet, and they thought they'd found the news. The increase went from 0.5 per cent – a figure that might be a gradual trend, but it could equally well be an entirely chance finding – to a front-page story in *The Times* about cocaine use doubling. You might not trust the press release, but if you don't know about numbers, then you take a big chance when you delve under the bonnet of a study to find a story.

OK, back to an easy one

There are also some perfectly simple ways to generate ridiculous statistics, and two common favourites are to select an unusual sample group, and to ask them a stupid question. Let's say 70 per cent of all women want Prince Charles to be told to stop interfering in public life. Oh, hang on – 70 per cent of all women *who visit my website* want Prince Charles to be told to stop interfering in public life. You can see where we're going. And of course, in surveys, if they are voluntary, there is something called *selection bias*: only the people who can be bothered to fill out the survey form will actually have a vote registered.

There was an excellent example of this in the *Telegraph* in the last days of 2007. 'Doctors Say No to Abortions in their Surgeries' was the headline. 'Family doctors are threatening a revolt against government plans to allow them to perform abortions in their surgeries, the *Daily Telegraph* can disclose.' A revolt? 'Four out of five GPs do not want to carry out terminations even though the idea is being tested in NHS pilot schemes, a survey has revealed.'

Where did these figures come from? A systematic survey of all GPs, with lots of chasing to catch the non-responders? Telephoning them at work? A postal survey, at least? No. It was an online vote on a doctors' chat site that produced this major news story. Here is the question, and the options given:

'GPs should carry out abortions in their surgeries'
Strongly agree, agree, don't know, disagree, strongly disagree.

We should be clear: I myself do not fully understand this question. Is that 'should' as in 'should'? As in 'ought to'? And in what circumstances? With extra training, time and money? With extra systems in place for adverse outcomes? And remember, this is a website where doctors – bless them – go to moan. Are they just saying no because they're grumbling about more work and low morale?

More than that, what exactly does 'abortion' mean here? Looking at the comments in the chat forum, I can tell you that plenty of the doctors seemed to think it was about surgical abortions, not the relatively safe oral pill for termination of pregnancy. Doctors aren't that bright, you see. Here are some quotes:

> This is a preposterous idea. How can GPs ever carry out abortions in their own surgeries. What if there was a major complication like uterine and bowel perforation?

> GP surgeries are the places par excellence where infective disorders present. The idea of undertaking there any sort of sterile procedure involving an abdominal organ is anathema.

> The only way it would or rather should happen is if GP practices have a surgical day care facility as part of their premises which is staffed by appropriately trained staff, i.e. theatre staff, anaesthetist and gynaecologist ... any surgical operation is not without its risks, and presumably [we] will undergo gynaecological surgical training in order to perform.

> What are we all going on about? Let's all carry out abortions in our surgeries, living rooms, kitchens, garages, corner shops, you know, just like in the old days.

And here's my favourite:

I think that the question is poorly worded and I hope that [the doctors' website] do not release the results of this poll to the *Daily Telegraph*.

Beating you up

It would be wrong to assume that the kinds of oversights we've covered so far are limited to the lower echelons of society, like doctors and journalists. Some of the most sobering examples come from the very top.

In 2006, after a major government report, the media reported that one murder a week is committed by someone with psychiatric problems. Psychiatrists should do better, the newspapers told us, and prevent more of these murders. All of us would agree, I'm sure, with any sensible measure to improve risk management and violence, and it's always timely to have a public debate about the ethics of detaining psychiatric patients (although in the name of fairness I'd like to see preventive detention discussed for all other potentially risky groups too – like alcoholics, the repeatedly violent, people who have abused staff in the job centre, and so on).

But to engage in this discussion, you need to understand the maths of predicting very rare events. Let's take a very concrete example, and look at the HIV test. What features of any diagnostic procedure do we measure in order to judge how useful it might be? Statisticians would say the blood test for HIV has a very high 'sensitivity', at 0.999. That means that if you do have the virus, there is a 99.9 per cent chance that the blood test will be positive. They would also say the test has a high 'specificity' of 0.9999 – so, if you are not infected, there is a 99.99 per cent chance that the test will be negative. What a smashing blood test.*

---

* The figures here are ballpark, from Gerd Gigerenzer's excellent book *Reckoning with Risk*.

But if you look at it from the perspective of the person being tested, the maths gets slightly counterintuitive. Because weirdly, the meaning, the predictive value, of an individual's positive or negative test is changed in different situations, depending on the background rarity of the event that the test is trying to detect. The rarer the event in your population, the worse your test becomes, even though it is the same test.

This is easier to understand with concrete figures. Let's say the HIV infection rate among high-risk men in a particular area is 1.5 per cent. We use our excellent blood test on 10,000 of these men, and we can expect 151 positive blood results overall: 150 will be our truly HIV-positive men, who will get true positive blood tests; and one will be the one false positive we could expect from having 10,000 HIV-negative men being given a test that is wrong one time in 10,000. So, if you get a positive HIV blood test result, in these circumstances your chances of being truly HIV positive are 150 out of 151. It's a highly predictive test.

Let's now use the same test where the background HIV infection rate in the population is about one in 10,000. If we test 10,000 people, we can expect two positive blood results overall. One from the person who really is HIV positive; and the one false positive that we could expect, again, from having 10,000 HIV-negative men being tested with a test that is wrong one time in 10,000.

Suddenly, when the background rate of an event is rare, even our previously brilliant blood test becomes a bit rubbish. For the two men with a positive HIV blood test result, in this population where only one in 10,000 has HIV, it's only 50:50 odds on whether they really are HIV positive.

Let's think about violence. The best predictive tool for psychiatric violence has a 'sensitivity' of 0.75, and a 'specificity' of 0.75. It's tougher to be accurate when predicting an event in humans, with human minds and changing human lives. Let's say 5 per cent of patients seen by a community mental health team will be involved in a violent event in a year. Using the same maths as we did for the HIV tests, your '0.75' predictive tool would be wrong eighty-six times out of a hundred. For serious violence, occurring at 1 per cent a year, with our best '0.75' tool, you inaccurately finger your potential perpetrator ninety-seven times out of a hundred. Will you preventively detain ninety-seven people to prevent three violent events? And will you apply that rule to alcoholics and assorted nasty antisocial types as well?

For murder, the extremely rare crime in question in this report, for which more action was demanded, occurring at one in 10,000 a year among patients with psychosis, the false positive rate is so high that the best predictive test is entirely useless.

This is not a counsel of despair. There are things that can be done, and you can always try to reduce the number of actual stark cock-ups, although it's difficult to know what proportion of the 'one murder a week' represents a clear failure of a system, since when you look back in history, through the retrospectoscope, anything that happens will look as if it was inexorably leading up to your one bad event. I'm just giving you the maths on rare events. What you do with it is a matter for you.

Locking you up

In 1999 solicitor Sally Clark was put on trial for murdering her two babies. Most people are aware that there was a statistical error in the prosecution case, but few know the true story, or the phenomenal extent of the statistical ignorance that went on in the case.

At her trial, Professor Sir Roy Meadow, an expert in parents who harm their children, was called to give expert evidence. Meadow famously quoted 'one in seventy-three million' as the chance of two children in the same family dying of Sudden Infant Death Syndrome (SIDS).

This was a very problematic piece of evidence for two very distinct reasons: one is easy to understand, the other is an absolute mindbender. Because you have the concentration span to follow the next two pages, you will come out smarter than Professor Sir Roy, the judge in the Sally Clark case, her defence teams, the appeal court judges, and almost all the journalists and legal commentators reporting on the case. We'll do the easy reason first.

The ecological fallacy

The figure of 'one in seventy-three million' itself is iffy, as everyone now accepts. It was calculated as 8,543 × 8,543, as if the chances of two SIDS episodes in this one family were independent of each other. This feels wrong from the outset, and anyone can see why: there might be environmental or genetic factors at play, both of which would be shared by the two babies. But forget how pleased you are with yourself for understanding that fact. Even if we accept that two SIDS in one family is much more likely than one in seventy-three million – say, one in 10,000 – any such figure is still of dubious relevance, as we will now see.

The prosecutor's fallacy

The real question in this case is: what do we do with this spurious number? Many press reports at the time stated that one in seventy-three million was the likelihood that the deaths of Sally Clark's two children were accidental: that is, the likelihood that she was innocent. Many in the court process seemed to share this view, and the factoid certainly sticks in the mind. But this is an example of a well-known and well-documented piece of flawed reasoning known as 'the prosecutor's fallacy'.

Two babies in one family have died. This in itself is very rare. Once this rare event has occurred, the jury needs to weigh up two competing explanations for the babies' deaths: double SIDS or double murder. Under normal circumstances – before any babies have died – double SIDS is very unlikely, and so is double murder. But now that the rare event of two babies dying in one family has occurred, the two explanations – double murder or double SIDS – are suddenly both very likely. If we really wanted to play statistics, we would need to know which is relatively *more* rare, double SIDS or double murder. People have tried to calculate the relative risks of these two events, and one paper says it comes out at around 2:1 in favour of double SIDS.

Not only was this *crucial* nuance of the prosecutor's fallacy missed at the time – by everyone in the court – it was also clearly missed in the appeal, at which the judges suggested that instead of 'one in seventy-three million', Meadow should have said 'very rare'. They recognised the flaws in its calculation, the ecological fallacy, the easy problem above, but they still accepted his number as establishing 'a very broad point, namely the rarity of double SIDS'.

That, as you now understand, was entirely wrongheaded: the rarity of double SIDS is irrelevant, because double murder is rare too. An entire court process failed to spot the nuance of how the figure should be used. Twice.

Meadow was foolish, and has been vilified (some might say this process was exacerbated by the witch-hunt against paediatricians who work on child abuse), but if it is true that he should have spotted and anticipated the problems in the interpretation of his number, then so should the rest of the people involved in the case: a paediatrician has no more unique responsibility to be numerate than a lawyer, a judge, journalist, jury member or clerk. The prosecutor's fallacy is also highly relevant in DNA evidence, for example, where interpretation frequently turns on complex mathematical and contextual issues. Anyone who is going to trade in numbers, and use them, and think with them, and persuade with them, let alone lock people up with them, also has a responsibility to understand them. All you've done is read a popular science book on them, and already you can see it's hardly rocket science.

Losing the lottery

You know, the most amazing thing happened to me tonight. I was coming here, on the way to the lecture, and I came in through the parking lot. And you won't believe what happened. I saw a car with the license plate ARW 357. Can you imagine? Of all the millions of license plates in the state, what was the chance that I would see that particular one tonight? Amazing …

Richard Feynman

It is possible to be very unlucky indeed. A nurse called Lucia de Berk has been in prison for six years in Holland, convicted of seven counts of murder and three of attempted murder. An unusually large number of people died when she was on shift, and that, essentially, along with some very weak circumstantial evidence, is the substance of the case against her. She has never confessed, she has continued to protest her innocence, and her trial has generated a small collection of theoretical papers in the statistics literature.

The judgement was largely based on a figure of 'one in 342 million against'. Even if we found errors in this figure – and believe me, we will – as in our previous story, the figure itself would still be largely irrelevant. Because, as we have already seen repeatedly, the interesting thing about statistics is not the tricky maths, but what the numbers mean.

There is also an important lesson here from which we could all benefit: unlikely things do happen. Somebody wins the lottery every week; children are struck by lightning. It's only weird and startling when something very, very specific and unlikely happens *if you have specifically predicted it beforehand.**

Here is an analogy.

Imagine I am standing near a large wooden barn with an enormous machine gun. I place a blindfold over my eyes and – laughing maniacally – I fire off many thousands and thousands of bullets into the side of the barn. I then drop the gun, walk over to the wall, examine it closely for some time, all over, pacing up and down. I find one spot where there are three bullet holes close to each other, then draw a target around them, announcing proudly that I am an excellent marksman.

You would, I think, disagree with both my methods and my conclusions for that deduction. But this is exactly what has happened in Lucia's case: the prosecutors found seven deaths, on one nurse's shifts, in one hospital, in one city, in one country, in the world, and then drew a target around them.

This breaks a cardinal rule of any research involving statistics: you cannot find your hypothesis in your results. Before you go to your data with your statistical tool, you have to have a specific hypothesis to test. If your hypothesis comes from analysing the data, then there is no sense in analysing the same data again to confirm it.

This is a rather complex, philosophical, mathematical form of circularity: but there were also very concrete forms of circular reasoning in the case. To collect more data, the investigators went back to the wards to see if they could find more suspicious deaths. But all the people who were asked to remember 'suspicious incidents' knew that they were being asked because Lucia might be a serial killer. There was a high risk that 'an incident

---

* The magician and pseudoscience debunker James Randi used to wake up every morning and write on a card in his pocket: 'I, James Randi, will die today', followed by the date and his signature. Just in case, he has recently explained, he really did, by some completely unpredictable accident.

was suspicious' became synonymous with 'Lucia was present'. Some sudden deaths when Lucia was not present would not be listed in the calculations, by definition: they are in no way suspicious, because Lucia was not present.

It gets worse. 'We were asked to make a list of incidents that happened during or shortly after Lucia's shifts,' said one hospital employee. In this manner more patterns were unearthed, and so it became even more likely that investigators would find more suspicious deaths on Lucia's shifts. Meanwhile, Lucia waited in prison for her trial.

This is the stuff of nightmares.

At the same time, a huge amount of corollary statistical information was almost completely ignored. In the three years before Lucia worked on the ward in question, there were seven deaths. In the three years that she did work on the ward, there were six deaths. Here's a thought: it seems odd that the death rate should go *down* on a ward at the precise moment that a serial killer – on a killing spree – arrives. If Lucia killed them all, then there must have been no natural deaths on that ward at all in the whole of the three years that she worked there.

Ah, but on the other hand, as the prosecution revealed at her trial, Lucia did like tarot. And she does sound a bit weird in her private diary, excerpts from which were read out. So she might have done it anyway.

But the strangest thing of all is this. In generating his obligatory, spurious, Meadowesque figure – which this time was 'one in 342 million' – the prosecution's statistician made a simple, rudimentary mathematical error. He combined individual statistical tests by multiplying p-values, the mathematical description of chance, or statistical significance. This bit's for the hardcore science nerds, and will be edited out by the publisher, but I intend to write it anyway: you do not just multiply p-values together, you weave them with a clever tool, like maybe 'Fisher's method for combination of independent p-values'.

If you multiply p-values together, then harmless and probable incidents rapidly appear vanishingly unlikely. Let's say you worked in twenty hospitals, each with a harmless incident pattern: say p=0.5. If you multiply those harmless p-values, of entirely chance findings, you end up with a final p-value of 0.5 to the power of twenty, which is $p < 0.000001$, which is extremely, very, highly statistically significant. With this mathematical error, by his reasoning, if you change hospitals a lot, you automatically become a suspect. Have you worked in twenty hospitals? For God's sake don't tell the Dutch police if you have.

14

Health Scares

In the previous chapter we looked at individual cases: they may have been egregious, and in some respects absurd, but the scope of the harm they can do is limited. We have already seen, with the example of Dr Spock's advice to parents on how their babies should sleep, that when your advice is followed by a very large number of people, if you are wrong, even with the best of intentions, you can do a great deal of harm, because the effects of modest tweaks in risk are magnified by the size of the population changing its behaviour.

It's for this reason that journalists have a special responsibility, and that's also why we will devote the last chapter of this book to examining the processes behind two very illustrative scare stories: the MRSA swabs hoax, and MMR. But as ever, as you know, we are talking about much more than just those two stories, and there will be many distractions along the way.

The Great MRSA Hoax

There are many ways in which journalists can mislead a reader with science: they can cherry-pick the evidence, or massage the statistics; they can pit hysteria and emotion against cold, bland statements from authority figures. The MRSA stings of 2005 come as close to simply 'making stuff up' as anything I've stumbled on so far.

I first worked out what was going on when I got a phone call from a friend who works as an undercover journalist for television. 'I just got a job as a cleaner to take some MRSA swabs for my *filthy hospital superbug scandal,*' he said, 'but they all came back negative. What am I doing wrong?' Happy to help, I explained that MRSA doesn't survive well on windows and doorknobs. The stories he had seen elsewhere were either rigged or rigged. Ten minutes later he rang back, triumphant: he had spoken to a health journalist from a well-known tabloid, and she had told him exactly which lab to use: 'the lab that always gives positive results' were the words she used, and it turned out to be Northants-based Chemsol Consulting, run by a man called Dr Christopher Malyszewicz. If you have ever seen an undercover MRSA superbug positive swab scandal, it definitely came from here. They all do.

Microbiologists at various hospitals had been baffled when their institutions fell victim to these stories. They took swabs from the same surfaces, and sent them to reputable mainstream labs, including their own: but the swabs grew nothing, contrary to Chemsol's results. An academic paper by eminent microbiologists describing this process in relation to one hospital – UCLH – was published in a peer-reviewed academic journal, and loudly ignored by everyone in the media.

Before we go any further, we should clarify one thing, and it relates to the whole of this section on health scares: it is very reasonable to worry about health risks, and to check them out carefully. Authorities are not to be trusted, and in this specific case, plenty of hospitals aren't as clean as we'd like them to be. Britain has more MRSA than many other countries, and this could be for any number of reasons, including infection control measures, cleanliness, prescribing patterns, or things we've not thought of yet (I'm talking off the top of my head here).

But we're looking at the issue of one private laboratory, with an awful lot of business from undercover journalists doing MRSA undercover swab stories, that seems to give an awful lot of positive results.

I decided to ring Dr Chris Malyszewicz and ask if he had any ideas why this should be.

He said he didn't know, and suggested that the hospital microbiologists might be taking swabs from the wrong places at the wrong times. They can often be incompetent, he explained. I asked why he thought the tabloids always chose his lab (producing almost twenty articles so far, including a memorable 'mop of death' front page in the *Sunday Mirror*). He had no idea. I asked why various microbiologists had said he refused to disclose his full methods, when they only wanted to replicate his techniques in their own labs in order to understand the discrepancy. He said he'd told them everything (I suspect in retrospect he was so confused that he believed this to be true). He also mispronounced the names of some very common bacteria.

It was at this point that I asked Dr Malyszewicz about his qualifications. I don't like to critique someone's work on the basis of who they are, but it felt like a fair question under the circumstances. On the telephone, being entirely straight, he just didn't feel like a man with the intellectual horsepower necessary to be running a complex microbiology laboratory.

He told me he had a BSc from Leicester University. Actually it's from Leicester Polytechnic. He told me he has a PhD. The *News of the World* called him 'respected MRSA specialist Dr Christopher Malyszewicz'. The *Sun* called him 'the UK's top MRSA expert' and 'microbiologist Christopher Malyszewicz'. He was similarly lauded in the *Evening Standard* and the *Daily Mirror*. On a hunch, I put a difficult question to him. He agreed that his was a 'non-accredited correspondence course PhD' from America. He agreed that his PhD was not recognised in the UK. He had no microbiology qualifications or training (as many journalists were repeatedly told by professional microbiologists). He was charming and very pleasant to talk to: eager to please. What was he doing in that lab?

There are lots of ways to distinguish one type of bacteria from another, and you can learn some of the tricks at home with a cheap toy microscope: you might look at them, to see what shape they are, or what kinds of dyes and stains they pick up. You can see what shapes and colours the colonies make as they grow on 'culture media' in a glass dish, and you can look at whether certain things in the culture media affect their growth (like the presence of certain antibiotics, or types of nutrient). Or you can do genetic fingerprinting on them. These are just a few examples.

I spoke to Dr Peter Wilson, a microbiologist at University College London who had tried to get some information from Dr Malyszewicz about his methods for detecting the presence of MRSA, but received only confusing half stories. He tried using batches of the growth media that Dr Malyszewicz was using, which he seemed to be relying on to distinguish MRSA from other species of bacteria, but it grew lots of things equally well. Then people started trying to get plates from Dr Malyszewicz which he claimed contained MRSA. He refused. Journalists were told about this. Finally he released eight plates. I spoke to the microbiologists who tested them.

On six of the eight, where Dr Malyszewicz PhD believed he had found MRSA, the lab found none at all (and these plates were subjected to meticulous and forensic microbiological analyses, including PCR, the technology behind 'genetic fingerprinting'). On two of the plates there was indeed MRSA; but it was a very unusual strain. Microbiologists have huge libraries of the genetic make-up of different types of infectious agents, which are used to survey how different diseases are travelling around the world. By using these banks we can see, for example, that a strain of the polio virus from Kano province in northern Nigeria, following their vaccine scare, has popped up killing people on the other side of the world (see page 277).

This strain of MRSA had never been found in any patient in the UK, and it had only ever been seen rarely in Australia. There is *very* little chance that this was found wild in the UK: in all likelihood it was a contaminant, from the media work ChemSol had done for Australian tabloids. On the other six plates, where Malyszewicz thought he had MRSA, there were mostly just bacilli, a common but completely different group of bacteria. MRSA looks like a ball. Bacilli look like a rod. You can tell the difference between them using 100x magnification – the 'Edu Science Microscope Set' at Toys'Я'Us for £9.99 will do the job very well (if you buy one, with the straightest face in the world, I recommend looking at your sperm: it's quite a soulful moment).

We can forgive journalists for not following the science detail. We can forgive them for being investigative newshounds, perhaps, even though they were repeatedly told – by perfectly normal microbiologists, not men in black – that Chemsol's results were improbable, and probably impossible. But was there anything else, something more concrete, that would have suggested to these journalists that their favourite lab was providing inaccurate results?

Perhaps yes, when they visited Malyszewicz's laboratory, which had none of the accreditation which you would expect for any normal lab. On just one occasion the government's Inspector of Microbiology was permitted to inspect it. The report from this visit describes the Chemsol laboratory as 'a freestanding, single storey wooden building, approximately 6m × 2m in the back garden'. It was a garden shed. They go on to describe 'benching of a good household quality (not to microbiology laboratory standards)'. It was a garden shed with kitchen fittings.

And we should also mention in passing that Malyszewicz had a commercial interest: 'Worried about MRSA? The perfect gift for a friend or relative in hospital. Show them how much you care for their health by giving a Combact Antimicrobial Hospital Pack. Making sure they come out fighting fit.' It turned out that most of Chemsol's money came from selling disinfectants for MRSA, often with bizarre promotional material.

How did the papers respond to the concerns, raised by senior microbiologists all over the country, that this man was providing bogus results? In July 2004, two days after Malyszewicz allowed these two real microbiologists in to examine his garden shed, the *Sunday Mirror* wrote a long, vitriolic piece about them: 'Health Secretary John Reid was accused last night of trying to gag Britain's leading expert on the killer bug MRSA.' Britain's leading expert who has no microbiology qualifications, runs his operation from a shed in the garden, mispronounces the names of common bacteria, and demonstrably doesn't understand the most basic aspects of microbiology. 'Dr Chris Malyszewicz has pioneered a new method of testing for levels of MRSA and other bacteria,' it went on. 'They asked me a lot of questions about my procedures and academic background,' said Dr Malyszewicz. 'It was an outrageous attempt to discredit and silence him,' said Tony Field, chairman of the national MRSA support group, who inevitably regarded Dr Malyszewicz as a hero, as did many who had suffered at the hands of this bacterium.

The accompanying editorial in the *Sunday Mirror* heroically managed to knit three all-time classic bogus science stories together, into one stirring eulogy:

> Whistle-blowers appear to bring out the very worst in this Government.
> NO WAY TO TREAT A DEDICATED DOCTOR
> First, Frankenstein foods expert Arpad Puzstai felt Labour's wrath when he dared to raise the alarm over genetically-modified

crops. Then Dr Andrew Wakefield suffered the same fate when he suggested a link between the single-jab MMR vaccine and autism. Now it's the turn of Dr Chris Malyszewicz, who has publicly exposed alarmingly high rates of the killer bug MRSA in NHS hospitals.

Dr Chris Malyszewicz should get a medal for his work. Instead he tells the *Sunday Mirror* how Health Secretary John Reid sent two senior advisers to his home to 'silence him'.

The *Sunday Mirror* was not alone. When the *Evening Standard* published an article based on Malyszewicz's results ('Killer Bugs Widespread in Horrifying Hospital Study'), two senior consultant microbiologists from UCH, Dr Geoff Ridgway and Dr Peter Wilson, wrote to the paper pointing out the problems with Malyszewicz's methods. The *Evening Standard* didn't bother to reply.

Two months later it ran another story using Malyszewicz's bogus results. That time Dr Vanya Gant, another UCH consultant microbiologist, wrote to the paper. This time the *Standard* did deign to reply:

> We stand by the accuracy and integrity of our articles. The research was carried out by a competent person using current testing media. Chris Malyszewicz … is a fully trained microbiologist with eighteen years' experience … We believe the test media used … were sufficient to detect the presence of pathogenic type MRSA.

What you are seeing here is a tabloid journalist telling a department of world-class research microbiologists that they are mistaken about microbiology. This is an excellent example of a phenomenon described in one of my favourite psychology papers: 'Unskilled and Unaware of It: How Difficulties in Recognizing One's Own Incompetence Lead to Inflated Self-

Assessments', by Justin Kruger and David Dunning. They noted that people who are incompetent suffer a dual burden: not only are they incompetent, but they may also be too incompetent to assay their own incompetence, because the skills which underlie an ability to *make* a correct judgement are the same as the skills required to *recognise* a correct judgement.

As has been noted, surveys repeatedly show that a majority of us consider ourselves to be above average at various skills, including leadership, getting on with other people, and expressing ourselves. More than that, previous studies had already found that unskilled readers are less able to rate their own text comprehension, bad drivers are poor at predicting their own performance on a reaction-time test, poorly performing students are worse at predicting test performance, and most chillingly, socially incompetent boys are essentially unaware of their repeated *faux pas.*

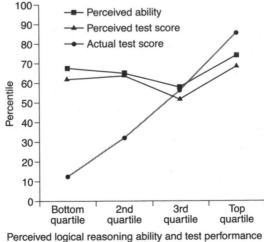

Perceived logical reasoning ability and test performance as a function of actual test performance

Kruger and Dunning brought this evidence together, but also did a series of new experiments themselves, looking at skills in domains like humour and logical reasoning. Their findings were twofold: people who performed particularly poorly relative to their peers were unaware of their own incompetence; but more than that, they were also less able to recognize competence *in others*, because this, too, relied on 'meta-cognition', or knowledge about the skill.

That was a pop-psych distraction. There is also a second, more general point to be made here. Journalists frequently flatter themselves with the fantasy that they are unveiling vast conspiracies, that the entire medical establishment has joined hands to suppress an awful truth. In reality I would guess that the 150,000 doctors in the UK could barely agree on second-line management of hypertension, but no matter: this fantasy was the structure of the MMR story, and the MRSA swab story, and many others, but it was a similar grandiosity that drove many of the earlier examples in this book where a journalist concluded that they knew best, including 'cocaine use doubles in the playground'.

Frequently, journalists will cite 'thalidomide' as if this was investigative journalism's greatest triumph in medicine, where they bravely exposed the risks of the drug in the face of medical indifference: it comes up almost every time I lecture on the media's crimes in science, and that is why I will explain the story in some detail here, because in reality – sadly, really – this finest hour never occurred.

In 1957, a baby was born with no ears to the wife of an employee at Grunenthal, the German drug company. He had taken their new anti-nausea drug home for his wife to try while she was pregnant, a full year before it went on the market: this is an illustration both of how slapdash things were, and of how difficult it is to spot a pattern from a single event.

The drug went to market, and between 1958 and 1962 around 10,000 children were born with severe malformations, all around the world, caused by this same drug, thalidomide. Because there was no central monitoring of malformations or adverse reactions, this pattern was missed. An Australian obstetrician called William McBride first raised the alarm in a medical journal, publishing a letter in the *Lancet* in December 1961. He ran a large obstetric unit, seeing a large number of cases, and he was rightly regarded as a hero – receiving a CBE – but it's sobering to think that he was only in such a good position to spot the pattern because he had prescribed so much of the drug, without knowing its risks, to his patients.* By the time his letter was published, a German paediatrician had noted a similar pattern, and the results of his study had been described in a German Sunday newspaper a few weeks earlier.

Almost immediately afterwards, the drug was taken off the market, and pharmacovigilance began in earnest, with notification schemes set up around the world, however imperfect you may find them to be. If you ever suspect that you've experienced an adverse drug reaction, as a member of the public, I would regard it as your duty to fill out a yellow card form online at yellowcard.mhra.gov.uk: anyone can do so. These reports can be collated and monitored as an early warning sign, and are a part of the imperfect, pragmatic monitoring system for picking up problems with medications.

No journalists were or are involved in this process. In fact Philip Knightley – a god of investigative journalism from the *Sunday Times'* legendary Insight team, and the man most associated with heroic coverage on thalidomide – specifically writes in his autobiography about his shame over not covering the thalidomide story sooner. They covered the political issue of compensation, rather well (it's more the *oeuvre* of journalists after all) but even that was done very late in the day, due to

---

* Many years later William McBride turned out to be guilty, in an unfortunate twist, of research fraud, falsifying data, and he was struck off the medical register in 1993, although he was later reinstated.

heinous legal threats from Grunenthal throughout the late 1960s and early 1970s.

Medical journalists, despite what they may try to tell you, most certainly did not reveal the dangers of thalidomide: and in many respects it's difficult to picture a world in which the characters who produce bogus MRSA hoax stories could somehow be meaningfully engaged in monitoring and administering drug safety, ably assisted, perhaps, by 'leading experts' from their garden sheds.

What the MRSA episode reveals to me, alongside a gut-wrenching and cavalier grandiosity, is the very same parody that we saw in our earlier review of nonsense science stories: humanities graduates in the media, perhaps feeling intellectually offended by how hard they find the science, conclude that it must simply be arbitrary, made up nonsense, to everyone. You can pick a result from anywhere you like, and if it suits your agenda, then that's that: nobody can take it away from you with their clever words, because it's all just game-playing, it just depends on who you ask, none of it really means anything, you don't understand the long words, and therefore, crucially, probably, *neither do the scientists*.

Epilogue

Although he was a very pleasant man, from my first telephone conversation with Chris Malyszewicz it was immediately clear that he lacked the basic background knowledge necessary to hold even a rudimentary discussion about microbiology. Patronising as it may sound, I feel a genuine sympathy for him, almost as a Walter Mitty figure. He claimed to have consulted for 'Cosworth-Technology, Boeing Aircraft, British Airways, Britannia Airways, Monarch Airways, Birmingham European Airways'. After BA and Boeing, neither of which had any record of any dealings with him, I gave up contacting these organisations. He would send elliptical comments in response to detailed criticisms of his 'analytic techniques', such as they were.

> Dear Ben,
>
> As a quote:
> 'I am surprised, but knowing what I know am not and knowing what I mean'.
>
> Thanks,
> Chris

I have strong feelings on this story: I do not blame Chris. I am certain that the true nature of his expertise would have been clear to anybody who spoke with him, regardless of background knowledge, and in my view it is the media that should have known better, with their huge offices, chains of command and responsibility, codes of conduct and editorial policies: not one man, in a shed in his back garden, surrounded by kitchen fittings and laboratory equipment he barely understood, bought on bank loans he was struggling to repay, in a small conurbation just outside Northampton.

Chris wasn't happy with what I wrote about him, and what was said about him after the story was exposed. We spent some time on the telephone, with him upset and me feeling, in all honesty, quite guilty. He felt that what was happening to him was unfair. He explained that he had never sought to be an expert on MRSA, but after the first story the journalists simply kept coming back, and everything snowballed. He may have made some mistakes, but he only wanted to help.

Chris Malyszewicz died in a car accident after losing control of his vehicle near Northampton shortly after the MRSA stories were exposed. He was heavily in debt.

15

The Media's MMR Hoax

The MRSA swab scandals were a simple, circumscribed, collective hoax. MMR is something much bigger: it is the prototypical health scare, by which all others must be judged and understood. It has every ingredient, every canard, every sleight of hand, and every aspect of venal incompetence and hysteria, systemic and individual. Even now, it is with great trepidation that I even dare to mention it by name, for two very simple reasons.

Firstly, at the quietest hint of a discussion on the subject, an army of campaigners and columnists will still, even in 2008, hammer on editors' doors demanding the right to a lengthy, misleading and emotive response in the name of 'balance'. Their demands are always, without exception, accommodated.

But there is a second issue, which is less important than it seems at first: Andrew Wakefield, the doctor who many imagine to be at the centre of the story, is currently in front of the GMC on charges of professional misconduct, and between me finishing and you reading this book, the judgement will probably be out.

I have no idea what that judgement will be, and being honest, although I suppose I'm glad they look into things like this in general, cases like his are two a penny at the GMC. I have no great interest in whether one individual's work was ethically dubious: the responsibility for the MMR scare cannot be laid at the door of a single man, however much the media may now be trying to argue that it should.

The blame lies instead with the hundreds of journalists, columnists, editors and executives who drove this story cynically, irrationally, and wilfully onto the front pages for nine solid years. As we will see, they overextrapolated from one study into absurdity, while studiously ignoring all reassuring data, and all subsequent refutations. They quoted 'experts' as authorities instead of explaining the science, they ignored the historical context, they set idiots to cover the facts, they pitched emotive stories from parents against bland academics (who they smeared), and most bizarrely of all, in some cases they simply made stuff up.

Now they claim that the original 1998 Wakefield research has been 'debunked' (it was never anything compelling in the first place), and you will be able to watch this year as they try to pin the whole scare onto one man. I'm a doctor too, and I don't imagine for one moment that I could stand up and create a nine-year-long news story on a whim. It is because of the media's blindness – and their unwillingness to accept their responsibility – that they will continue to commit the same crimes in the future. There is nothing you can do about that, so it might be worth paying attention now.

To remind ourselves, here is the story of MMR as it appeared in the British news media from 1998 onwards:

- Autism is becoming more common, although nobody knows why.
- A doctor called Andrew Wakefield has done scientific research showing a link between the MMR triple jab and autism.
- Since then, more scientific research has been done confirming this link.

- There is evidence that single jabs might be safer, but government doctors and those in the pay of the pharmaceutical industry have simply rubbished these claims.
- Tony Blair probably didn't give his young son the vaccine.
- Measles isn't so bad.
- And vaccination didn't prevent it very well anyway.

I think that's pretty fair. The central claim for each of these bullet points was either misleading or downright untrue, as we will see.

Vaccine scares in context

Before we begin, it's worth taking a moment to look at vaccine scares around the world, because I'm always struck by how circumscribed these panics are, and how poorly they propagate themselves in different soils. The MMR and autism scare, for example, is practically non-existent outside Britain, even in Europe and America. But throughout the 1990s France was in the grip of a scare that hepatitis B vaccine caused multiple sclerosis (it wouldn't surprise me if I was the first person to tell you that).

In the US, the major vaccine fear has been around the use of a preservative called thiomersal, although somehow this hasn't caught on here, even though that same preservative was used in Britain. And in the 1970s – since the past is another country too – there was a widespread concern in the UK, driven again by a single doctor, that whooping-cough vaccine was causing neurological damage.

Looking even further back, there was a strong anti-smallpox-vaccine movement in Leicester well into the 1930s, despite its demonstrable benefits, and in fact anti-inoculation sentiment goes right back to its origins: when James Jurin studied inoculation against smallpox (finding that it was associated with a lower death rate than the natural disease), his newfangled numbers and statistical ideas were treated with enormous suspicion. Indeed, smallpox inoculation remained illegal in France until 1769.* Even when Edward Jenner introduced the much safer vaccination for protecting people against smallpox at the turn of the nineteenth century, he was strongly opposed by the London cognoscenti.

And in an article from *Scientific American* in 1888 you can find the very same arguments which modern antivaccination campaigners continue to use today:

> The success of the anti-vaccinationists has been aptly shown by the results in Zurich, Switzerland, where for a number of years, until 1883, a compulsory vaccination law obtained, and small-pox was wholly prevented – not a single case occurred in 1882. This result was seized upon the following year by the anti-vaccinationists and used against the necessity for any such law, and it seems they had sufficient influence to cause its repeal. The death returns for that year (1883) showed that for every 1,000 deaths two were caused by smallpox; In 1884 there were three; in 1885, 17, and in the first quarter of 1886, 85.

Meanwhile, WHO's highly successful global polio eradication programme was on target to have eradicated this murderous

---

* Disdain for statistics in healthcare research wasn't unusual at the time: Ignaz Semmelweis noticed in 1847 that patients were dying much more frequently on the obstetrics ward run by the medical students than on the one run by the midwifery students (this was in the days when students did all the legwork in hospitals). He was pretty sure that this was because the medical students were carrying something nasty from the corpses in the dissection room, so he instituted proper handwashing practices with chlorinated lime, and did some figures on the benefits. The death rates fell, but in an era of medicine that championed 'theory' over real-world empirical evidence, he was basically ignored, until Louis Pasteur came along and confirmed the germ theory. Semmelweis died alone in an asylum. You've heard of Pasteur.

disease from the face of the earth by now – a fate which has already befallen the smallpox virus, excepting a few glass vials – until local imams from a small province called Kano in northern Nigeria claimed that the vaccine was part of a US plot to spread AIDS and infertility in the Islamic world, and organised a boycott which rapidly spread to five other states in the country. This was followed by a large outbreak of polio in Nigeria and surrounding countries, and tragically even further afield. There have now been outbreaks in Yemen and Indonesia, causing lifelong paralysis in children, and laboratory analysis of the genetic code has shown that these outbreaks were caused by the same strain of the polio virus, exported from Kano.

After all, as any trendy MMR-dodging north-London middle class humanities-graduate couple with children would agree, just because vaccination has almost eradicated polio – a debilitating disease which as recently as 1988 was endemic in 125 countries – that doesn't necessarily mean it's a good thing.

The diversity and isolation of these anti-vaccination panics helps to illustrate the way in which they reflect local political and social concerns more than a genuine appraisal of the risk data: because if the vaccine for hepatitis B, or MMR, or polio, is dangerous in one country, it should be equally dangerous everywhere on the planet; and if those concerns were genuinely grounded in the evidence, especially in an age of the rapid propagation of information, you would expect the concerns to be expressed by journalists everywhere. They're not.

Andrew Wakefield and his Lancet paper

In February 1998 a group of researchers and doctors led by a surgeon called Andrew Wakefield from the Royal Free Hospital in London published a research paper in the *Lancet* which by now stands as one of the most misunderstood and misreported papers in the history of academia. In some respects it did itself no favours: it is badly written, and has no clear statement of its hypothesis, or indeed of its conclusions (you can read it free online if you like). It has since been partially retracted.

The paper described twelve children who had bowel problems and behavioural problems (mostly autism), and mentioned that the parents or doctors of eight of these children believed that their child's problems had started within a few days of them being given the MMR vaccine. It also reported various blood tests, and tests on tissue samples taken from the children. The results of these were sometimes abnormal, but varied between children.

> 12 children, consecutively referred to the department of paediatric gastroenterology with a history of a pervasive developmental disorder with loss of acquired skills and intestinal symptoms (diarrhoea, abdominal pain, bloating and food intolerance), were investigated.
>
> … In eight children, the onset of behavioural problems had been linked, either by the parents or by the child's physician, with measles, mumps, and rubella vaccination … In these eight children the average interval from exposure to first behavioural symptoms was 6.3 days (range 1–14).

What can this kind of paper tell you about a link between something as common as MMR, and something as common as autism? Basically nothing, either way. It was a collection of twelve clinical anecdotes, a type of paper called a 'case series' – and a case series, by design, wouldn't demonstrate such a relationship between an exposure and an outcome with any force. It did not take some children who were given MMR and some children who weren't, and then compare the rates of autism between the two groups (this would have been a 'cohort study'). It did not take some children with autism, and some children without autism, and then compare the rates of vaccination between the two groups (this would have been a 'case-control study').

Could anything else explain the apparent connection between MMR, bowel problems and autism in these eight children? Firstly, although they sound like rare things to come together, this was a specialist centre in a teaching hospital, and the children had only been referred there because they had bowel problems and behavioural problems (the circumstances of these referrals are currently being examined by the GMC, as we will see).

Out of an entire nation of millions of inhabitants, if some children with a combination of fairly common things (vaccination, autism, bowel problems) all come together in one place which is *already acting as beacon for such a combination*, as this clinic was, we should not naturally be impressed. You will remember from the discussion of the unfortunate Dutch nurse Lucia de Berk (and indeed from reading news reports about lottery winners) that unlikely combinations of events will always happen, somewhere, to some people, entirely by chance. Drawing a target around them after the fact tells us nothing at all.

All stories about treatment and risk will start with modest clinical hunches like these anecdotes; but hunches, with nothing to back them up, are not generally newsworthy. At the publication of this paper, a press conference was held at the Royal Free Hospital, and to the visible surprise of many other clinicians and academics present, Andrew Wakefield announced that he thought it would be prudent to use single vaccines instead of the MMR triple vaccine. Nobody should have been surprised: a video news release had already been issued by the hospital, in which Wakefield made the same call.

We are all entitled to our clinical hunches, as individuals, but there was nothing in either this study of twelve children, or any other published research, to suggest that giving single vaccines would be safer. As it happens, there are good grounds for believing that giving vaccines separately might be more harmful: they need six visits to the GP, and six unpleasant jabs, which makes four more appointments to miss. Maybe you're ill, maybe you're on holiday, maybe you move house, maybe you lose track of which ones you've had, maybe you can't see the point of rubella for boys, or mumps for girls, or maybe you're a working single mum with two kids and no time.

Also, of course, the children spend much more time vulnerable to infection, especially if you wait a year between jabs, as Wakefield has recommended, out of the blue. Ironically, although most of the causes of autism remain unclear, one of the few well-characterised single causes is rubella infection itself, while the child is in the womb.

The story behind the paper

Some fairly worrying questions have been raised since then. We won't cover them in detail, because I don't find *ad hominem* stories very interesting to write about, and because I don't want that aspect of the story – rather than the research evidence – to be the reason why you come to your own conclusion about the risks of MMR and autism. There are things which came out in 2004, however, which cannot fairly be ignored, including allegations of multiple conflicts of interest, undeclared sources of bias in the recruitment of subjects for the paper, undisclosed negative findings, and problems with the ethical clearance for the tests. These were largely uncovered by a tenacious investigative journalist from the *Sunday Times* called Brian Deer, and they now form part of the allegations being investigated by the GMC.

For example, it is investigating whether Wakefield failed to disclose to the editor of the *Lancet* his involvement in a patent relating to a new vaccine; more worrying are the concerns about where the twelve children in the 1998 Royal Free study came from. While in the paper it is stated that they were sequential referrals to a clinic, in fact Wakefield was already being paid £50,000 of legal aid money by a firm of solicitors to investigate children whose parents were preparing a case against MMR, and the GMC is further investigating where the patients in the study came from, because it seems that many of Wakefield's referrals had come to him specifically as someone who could show a link between MMR and autism, whether formally or informally, and was working on a legal case. This is the beacon problem once more, and under these circumstances, the fact that *only* eight of the twelve children's parents or physicians believed the problems were caused by MMR would be unimpressive, if anything.

Of the twelve children in the paper, eleven sued drug companies (the one that didn't was American), and ten of them already had legal aid to sue over MMR before the 1998 paper was published. Wakefield himself eventually received £435,643 plus expenses from the legal aid fund for his role in the case against MMR.

Various intrusive clinical investigations – such as lumbar punctures and colonoscopies – were carried out on the children, and these required ethics committee clearance. The Ethics Committee had been assured that they were all clinically indicated, which is to say, in the interests of the children's own clinical care: the GMC is now examining whether they were contrary to the clinical interests of the children, and performed simply for research.

Lumbar puncture involves putting a needle into the centre of the spine to tap off some spinal fluid, and colonoscopy involves putting a flexible camera and light through the anus, up the rectum and into the bowel on a long tube. Neither is without risk, and indeed one of the children being investigated as part of an extension of the MMR research project was seriously harmed during colonoscopy, and was rushed to intensive care at Great Ormond Street Hospital after his bowel was punctured in twelve places. He suffered multiple organ failure, including kidney and liver problems, and neurological injuries, and received £482,300 in compensation. These things happen, nobody is to blame, and I am merely illustrating the reasons to be cautious about doing investigations.

Meanwhile, in 1997 a young PhD student called Nick Chadwick was starting his research career in Andrew Wakefield's lab, using PCR technology (used as part of DNA fingerprinting) to look for traces of measles strain genetic material in the bowels of these twelve children, because this was a central feature of Wakefield's theory. In 2004 Chadwick gave an interview to Channel 4's *Dispatches*, and in 2006 he gave evidence at a US case on vaccines, stating that there was no measles DNA to be found in these samples. But this important finding, which conflicted with his charismatic supervisor's theory, was not published.

I could go on.

Nobody knew about any of this in 1998. In any case, it's not relevant, because the greatest tragedy of the media's MMR hoax is that it was brought to an end by these issues being made public, when it should have been terminated by a cautious and balanced appraisal of the evidence at the time. Now, you will see news reporters – including the BBC – saying stupid things like 'The research has since been debunked.' Wrong. The research never justified the media's ludicrous over-interpretation. If they had paid attention, the scare would never have even started.

The press coverage begins

What's most striking about the MMR scare – and this is often forgotten – is that it didn't actually begin in 1998. The *Guardian* and the *Independent* covered the press conference on their front pages, but the *Sun* ignored it entirely, and the *Daily Mail*, international journal of health scares, buried their piece on it in the middle of the paper. Coverage of the story was generally written by specialist health and science journalists, and they were often fairly capable of balancing the risks and evidence. The story was pretty soft.

In 2001 the scare began to gain momentum. Wakefield published a review paper in an obscure journal, questioning the safety of the immunisation programme, although with no new evidence. In March he published new laboratory work with Japanese researchers ('the Kawashima paper'), using PCR data to show measles virus in the white blood cells of children with bowel problems and autism. This was essentially the opposite of the findings from Nick Chadwick in Wakefield's own labs. Chadwick's work remained unmentioned (and there has since been a paper published showing how the Kawashima paper produced a false positive, although the media completely ignored this development, and Wakefield seems to have withdrawn his support for the study).

Things began to deteriorate. The anti-vaccination campaigners began to roll their formidable and coordinated publicity machine into action against a rather chaotic shambles of independent doctors from various different uncoordinated agencies. Emotive anecdotes from distressed parents were pitted against old duffers in corduroy, with no media training, talking about scientific data. If you ever wanted to see evidence against the existence of a sinister medical conspiracy, you need look no further than the shower of avoidant doctors and academics, and their piecemeal engagement with the media during this time. The Royal College of General Practitioners not only failed to speak clearly on the evidence, it also managed – heroically – to dig up some anti-MMR GPs to offer to journalists when they rang in asking for quotes.

The story began to gain momentum, perhaps bound up in the wider desire of some newspapers and personalities simply to attack the government and the health service. A stance on MMR became part of many newspapers' editorial policies, and that stance was often bound up with rumours about senior managerial figures with family members who had been affected by autism. It was the perfect story, with a single charismatic maverick fighting against the system, a Galileo-like figure; there were elements of risk, of awful personal tragedy, and of course, the question of blame. Whose fault was autism? Because nestling in the background was this extraordinary new diagnosis, a disease that struck down young boys and seemed to have come out of the blue, without explanation.

Autism

We still don't know what causes autism. A history of psychiatric problems in the family, early birth, problems at birth, and breech presentation are all risk factors, but pretty modest ones, which means they're interesting from a research perspective, but none of them explains the condition in a particular person. This is often the case with risk factors. Boys are affected more than girls, and the incidence of autism continues to rise, in part because of improved diagnosis – people who were previously given labels like 'mentally subnormal' or 'schizophrenia' were now receiving a diagnosis of 'autism' – but also possibly because of other factors which are still not understood. Into this vacuum of uncertainty, the MMR story appeared.

There was also something strangely attractive about autism as an idea to journalists and other commentators. Among other things, it's a disorder of language, which might touch a particular chord with writers; but it's also philosophically enjoyable to think about, because the flaws in social reasoning which are exhibited by people with autism give us an excuse to talk and think about our social norms and conventions. Books about autism and the autistic outlook on the world have become bestsellers. Here are some wise words for us all from Luke Jackson, a thirteen-year-old with Asperger's syndrome, who has written a book of advice for teenagers with the condition (*Freaks, Geeks and Asperger Syndrome*). This is from the section on dating:

If the person asks something like 'Does my bum look fat?' or even 'I am not sure I like this dress' then that is called 'fishing for compliments'. These are very hard things to understand, but I am told that instead of being completely honest and saying that yes their bum does look fat, it is politer to answer with something like 'Don't be daft, you look great.' You are not lying, simply evading an awkward question and complimenting them at the same time. Be economical with the truth!

Asperger's syndrome, or autistic spectrum disorder, is being applied to an increasingly large number of people, and children or adults who might previously have been considered 'quirky' now frequently have their personality medicalised with suggestions that they have 'traits of Asperger's'. Its growth as a pseudo-diagnostic category has taken on similar proportions to 'mild dyslexia' – you will have your own views on whether this process is helpful – and its widespread use has allowed us all to feel that we can participate in the wonder and mystery of autism, each with a personal connection to the MMR scare.

Except of course, in most cases, genuine autism is a pervasive developmental disorder, and most people with autism don't write quirky books about their odd take on the world which reveal so much to us about our conventions and social mores in a charmingly plain and unselfconscious narrative style. Similarly, most people with autism do not have the telegenic single skills which the media have so enjoyed talking up in their crass documentaries, like being *really amazing* at mental arithmetic, or playing the piano to concert standard while staring confusedly into the middle distance.

That these are the sort of things most people think of when the word 'autism' pops into their head is testament to the mythologisation and paradoxical 'popularity' of the diagnosis. Mike Fitzpatrick, a GP with a son who has autism, says that there are two questions on the subject which will make him want to slap you. One is: 'Do you think it was caused by MMR?' The other is: 'Does he have any special skills?'

Leo Blair

But the biggest public health disaster of all was a sweet little baby called Leo. In December 2001 the Blairs were asked if their infant son had been given the MMR vaccine, and refused to answer. Most other politicians have been happy to clarify whether their children have had the vaccine, but you can imagine how people might believe the Blairs were the kind of family not to have their children immunised, especially with everyone talking about 'herd immunity', and the worry that they might be immunising their child, and placing it at risk, in order that the rest of the population should be safer.

Concerns were particularly raised by the ubiquity of Cherie Blair's closest friend and aide. Carole Caplin was a New Age guru, a 'life coach' and a 'people person', although her boyfriend, Peter Foster, was a convicted fraudster. Foster helped arrange the Blairs' property deals, and he also says that they took Leo to a New Age healer, Jack Temple, who offered crystal dowsing, homoeopathy, herbalism and neolithic-circle healing in his back garden.

I'm not sure how much credence to give to Foster's claims myself, but the impact on the MMR scare is that they were widely reported at the time. We were told that the Prime Minister of the United Kingdom agreed to Temple waving a crystal pendulum over his son to protect him (and therefore his classmates, of course) from measles, mumps and rubella; and that Tony let Cherie give Temple some of his own hair and nail clippings, which Temple preserved in jars of alcohol. He said he only needed to swing his pendulum over the jar to know if their owner was healthy or ill.

Some things are certainly true. Using this crystal dowsing pendulum, Temple did claim that he could harness energy from heavenly bodies. He sold remedies with names like 'Volcanic Memory', 'Rancid Butter', 'Monkey Sticks', 'Banana Stem' and, my own personal favourite, 'Sphincter'. He was also a very well-connected man. Jerry Hall endorsed him. The Duchess of York wrote the introduction to his book *The Healer: The Extraordinary Healing Methods of Jack Temple* (it's a hoot). He told the *Daily Mail* that babies who are breastfed from the moment of birth acquire natural immunity against all diseases, and he even sold a homoeopathic alternative to the MMR jab.

> 'I tell all my patients who are pregnant that when the baby is born they must put it on the breast until there is no longer a pulse in the umbilical cord. It usually takes about 30 minutes. By doing this they transfer the mother's immune system to the baby, who will then have a fully-functioning immune system and will not need vaccines.' ... Mr Temple refused to confirm yesterday whether he advised Mrs Blair not to have her baby Leo vaccinated. But he said: 'If women follow my advice their children will not need the MMR injection, end of story.'*
>
> Daily Mail, *26 December 2001*

Cherie Blair was also a regular visitor to Carole's mum, Sylvia Caplin, a spiritual guru. 'There was a particularly active period in the summer when Sylvia was channelling for Cherie over two or three times a week, with almost daily contact between them,' the *Mail* reported. 'There were times when Cherie's faxes ran to

---

* Here is Jack on cramp: 'For years many people have suffered with cramp. By dowsing, I discovered that this is due to the fact that the body is not absorbing the element "scandium" which is linked to and controls the absorption of magnesium phosphate.' And on general health complaints: 'Based on my expertise in dowsing, I noted that many of my patients were suffering from severe deficiencies of carbon in their systems. The ease in which people these days suffer hairline fractures and broken bones is glaringly apparent to the eyes that are trained to see.'

10 pages.' Sylvia, along with many if not most alternative therapists, was viciously anti-MMR (over half of all the homeopaths
approached in one survey grandly advised against the vaccine).
The *Daily Telegraph* reported:

> We move on to what is potentially a very political subject: the
> MMR vaccine. The Blairs publicly endorsed it, then caused a
> minor furore by refusing to say whether their baby, Leo, had been
> inoculated. Sylvia [Caplin] doesn't hesitate: 'I'm against it,' she
> says. 'I'm appalled at so much being given to little children. The
> thing about these drugs is the toxic substance they put the
> vaccines in – for a tiny child, the MMR is a ridiculous thing to do.
>
> 'It has definitely caused autism. All the denials that come
> from the old school of medicine are open to question because
> logic and common sense must tell you that there's some toxic
> substance in it. Do you not think that's going to have an effect
> on a tiny child? Would you allow it? No – too much, too soon, in
> the wrong formula.'

It was also reported – doubtless as part of a cheap smear – that
Cherie Blair and Carole Caplin encouraged the Prime Minister
to have Sylvia 'douse and consult The Light, believed by Sylvia
to be a higher being or God, by use of her pendulum' to decide
if it was safe to go to war in Iraq. And while we're on the subject,
in December 2001 *The Times* described the Blairs' holiday in
Temazcal, Mexico, where they rubbed fruits and mud over each
other's bodies inside a large pyramid on the beach, then
screamed while going through a New Age rebirthing ritual.
Then they made a wish for world peace.

I'm not saying I buy all of this. I'm just saying, this is what
people were thinking about when the Blairs refused to publicly
clarify the issue of whether they had given their child the MMR
vaccine as all hell broke loose around it. This is not a hunch.
Thirty-two per cent of all the stories written that year about

MMR mentioned whether Leo Blair had had the vaccine or not (even Andrew Wakefield was only mentioned in 25 per cent), and it was one of the most well-recalled facts about the story in population surveys. The public, quite understandably, were taking Leo Blair's treatment as a yardstick of the Prime Minister's confidence in the vaccine, and few could understand why it should be a secret, if it wasn't an issue.

The Blairs, meanwhile, cited their child's right to privacy, which they felt was more important than an emerging public health crisis. It's striking that Cherie Blair has now decided, in marketing her lucrative autobiography, to waive that principle which was so vital at the time, and has written at length in her heavily promoted book not just about the precise bonk that conceived Leo, but also about whether he had the jab (she says yes, but she seems to obfuscate on whether it was single vaccines, and indeed on the question of when he had it: frankly, I give up on these people).

For all that it may seem trite and voyeuristic to you, this event was central to the coverage of MMR. 2002 was the year of Leo Blair, the year of Wakefield's departure from the Royal Free, and it was the peak of the media coverage, by a very long margin.

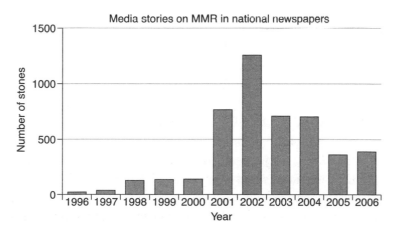

What was in these stories?

The MMR scare has created a small cottage industry of media analysis, so there is a fair amount known about the coverage. In 2003 the Economic and Social Research Council (ESRC) published a paper on the media's role in the public understanding of science, which sampled all the major science media stories from January to September 2002, the peak of the scare. Ten per cent of all science stories were about MMR, and MMR was also by far the most likely to generate letters to the press (so people were clearly engaging with the issue); by far the most likely science topic to be written about in opinion or editorial pieces; and it generated the longest stories. MMR was the biggest, most heavily covered science story for years.

Pieces on GM food, or cloning, stood a good chance of being written by specialist science reporters, but for stories on MMR these reporters were largely sidelined, and 80 per cent of the coverage of the biggest science story of the year was by generalist reporters. Suddenly we were getting comment and advice on complex matters of immunology and epidemiology from people who would more usually have been telling us about a funny thing that happened with the au pair on the way to a dinner party. Nigella Lawson, Libby Purves, Suzanne Moore, Lynda Lee-Potter and Carol Vorderman, to name only a few, all wrote about their ill-informed concerns on MMR, blowing hard on their toy trumpets. The anti-MMR lobby, meanwhile, developed a reputation for targeting generalist journalists wherever possible, feeding them stories, and actively avoiding health or science correspondents.

This is a pattern which has been seen before. If there is one thing which has adversely affected communication between scientists, journalists and the public, it is the fact that science journalists simply do not cover major science news stories. From drinking with science journalists, I know that much of the time, nobody even runs these major stories by them for a quick check.

Again, I'm not speaking in generalities here. During the crucial two days after the GM 'Frankenstein foods' story broke in February 1999, *not a single one* of the news articles, opinion pieces or editorials on the subject was written by a science journalist. A science correspondent would have told his or her editor that when someone presents their scientific findings about GM potatoes causing cancer in rats, as Arpad Pusztai did, on ITV's *World in Action* rather than in an academic journal, then there's something fishy going on. Pusztai's experiment was finally published a year later – after a long period when nobody could comment on it, because nobody knew what he'd actually done – and when all was revealed in a proper publication, his experimental results did not contain information to justify the media's scare.

This sidelining of specialist correspondents when science becomes front-page news, and the fact that they are not even used as a resource during these periods, has predictable consequences. Journalists are used to listening with a critical ear to briefings from press officers, politicians, PR executives, sales-people, lobbyists, celebrities and gossip-mongers, and they generally display a healthy natural scepticism: but in the case of science, they don't have the skills to critically appraise a piece of scientific evidence on its merits. At best the evidence of these 'experts' will only be examined in terms of who they are as people, or perhaps who they have worked for. Journalists – and many campaigners – think that this is what it means to critically appraise a scientific argument, and seem rather proud of themselves when they do it.

The scientific content of stories – the actual experimental evidence – is brushed over and replaced with didactic statements from authority figures on either side of the debate, which contributes to a pervasive sense that scientific advice is somehow arbitrary, and predicated upon a social role – the 'expert' – rather than on transparent and readily understandable empirical evidence. Worse than this, other elements are brought into the foreground: political issues, Tony Blair's refusal to say whether his baby had received the vaccine, mythical narratives, a lionised 'maverick' scientist, and emotive appeals from parents.

A reasonable member of the public, primed with such a compelling battery of human narrative, would be perfectly entitled to regard any expert who claimed MMR was safe as thoughtless and dismissive, especially if that claim came without any apparent supporting evidence.

The story was also compelling because, like GM food, the MMR story seemed to fit a fairly simple moral template, and one which I myself would subscribe to: big corporations are often dodgy, and politicians are not to be trusted. But it matters whether your political and moral hunches are carried in the right vehicle. Speaking only for myself, I am very wary of drug companies, not because I think all medicine is bad, but because I know they have hidden unflattering data, and because I have seen their promotional material misrepresent science. I also happen to be very wary of GM food – but not because of any inherent flaws in the technology, and not because I think it is uniquely dangerous. Somewhere between splicing in genes for products that will treat haemophilia at one end, and releasing genes for antibiotic resistance into the wild at the other, lies a sensible middle path for the regulation of GM, but there's nothing desperately remarkable or uniquely dangerous about it as a technology.

Despite all that, I remain extremely wary of GM for reasons that have nothing to do with the science, simply because it has created a dangerous power shift in agriculture, and 'terminator seeds', which die at the end of the season, are a way to increase farmers' dependency, both nationally and in the developing world, while placing the global food supply in the hands of multinational corporations. If you really want to dig deeper, Monsanto is also very simply an unpleasant company (it made Agent Orange during the Vietnam War, for example).

Witnessing the blind, seething, thoughtless campaigns against MMR and GM – which mirror the infantile train of thought that 'homeopathy works because the Vioxx side-effects were covered up by Merck' – it's easy to experience a pervasive sense of lost political opportunities, that somehow all of our valuable indignation about development issues, the role of big money in our society, and frank corporate malpractice, is being diverted away from anywhere it could be valid and useful, and into puerile, mythical fantasies. It strikes me that if you genuinely care about big business, the environment and health, then you're wasting your time with jokers like Pusztai and Wakefield.

Science coverage is further crippled, of course, by the fact that the subject can be quite difficult to understand. This in itself can seem like an insult to intelligent people, like journalists, who fancy themselves as able to understand most things, but there has also been an acceleration in complexity in recent times. Fifty years ago you could sketch out a full explanation of how an AM radio worked on the back of a napkin, using basic school-level knowledge of science, and build a crystal set in a classroom which was essentially the same as the one in your car. When your parents were young they could fix their own car, and understand the science behind most of the everyday technology they encountered, but this is no longer the case. Even a geek today would struggle to give an explanation of how his mobile phone works, because technology has become more difficult to understand and explain, and everyday gadgets have taken on a 'black box' complexity that can feel sinister, as well as intellectually undermining. The seeds were sown.

But we should return to the point. If there was little science, then what *did* appear in all these long stories on MMR? Going back to the 2002 data from the ESRC, only a quarter mentioned Andrew Wakefield, which seems odd, considering he was the cornerstone of the story. This created the erroneous impression that there was a large body of medical opinion which was suspicious of MMR, rather than just one 'maverick'. Less than a third of broadsheet reports referred to the overwhelming evidence that MMR is safe, and only 11 per cent mentioned that it is regarded as safe in the ninety other countries in which it is used.

It was rare to find much discussion of the evidence at all, as it was considered too complicated, and when doctors tried to explain it they were frequently shouted down, or worse still, their explanations were condensed into bland statements that 'science had shown' there was nothing to worry about. This uninformative dismissal was pitted against the emotive concerns of distressed parents.

As 2002 wore on, things got really strange. Some newspapers, such as the *Daily Mail* and the *Daily Telegraph*, made MMR the focus of a massive political campaign, and the beatification of Wakefield reached a kind of fever pitch. Lorraine Fraser had an exclusive interview with him in the *Telegraph* in which he was described as 'a champion of patients who feel their fears have been ignored'. She wrote a dozen similar articles over the next year (and her reward came when she was named British Press Awards Health Writer of the Year 2002, a gong I do not myself expect to receive).

Justine Picardie did a lavish photo feature on Wakefield, his house and his family for the *Telegraph* Saturday magazine. Andy is, she tells us, 'a handsome, glossy-haired hero to families of autistic children'. How are the family? 'A likeable, lively family, the kind you would be happy to have as friends, pitted against mysterious forces who have planted bugging devices and have stolen patients' records in "apparently inexplicable" burglaries.' She fantasises – and I absolutely promise you I'm not making this up – about a Hollywood depiction of Wakefield's heroic struggle, with Russell Crowe playing the lead 'opposite Julia Roberts as a feisty single mother fighting for justice for her child'.

The evidence on MMR

So what is the evidence on the safety of MMR?

There are a number of ways to approach the evidence on the safety of a given intervention, depending on how much attention you have to give. The simplest approach is to pick an arbitrary authority figure: a doctor, perhaps, although this seems not to be appealing (in surveys people say they trust doctors the most, and journalists the least: this shows the flaw in that kind of survey).

You could take another, larger authority at face value, if there is one that suits you. The Institute of Medicine, the Royal Colleges, the NHS, and more, all came out in support of MMR, but this was apparently not sufficient to convince. You could offer information: an NHS website at mmrthefacts.nhs.uk started with the phrase 'MMR is safe' (literally), and allowed the reader to drill down to the detail of individual studies.* But that too did little to stem the tide. Once a scare is running, perhaps every refutation can seem like an admission of guilt, drawing attention to the scare.

The Cochrane Collaboration is as blemishless as they come, and it has done a systematic review of the literature on MMR, concluding that there was no evidence that it is unsafe (although it didn't appear until 2005). This reviewed the data the media had systematically ignored: what was in it?

---

* Whether you buy the DoH phrase 'MMR is safe' depends on what you decide you mean by 'safe'. Is flying safe? Is your washing machine safe? What are you sitting on? Is that safe? You can obsess over the idea that philosophically nothing can ever be shown to be 100 per cent safe – and many will – but you would be arguing about a fairly meaningless and uncommon definition of the word.

If we are to maintain the moral high ground, there are a few things we need to understand about evidence. Firstly, there is no single golden study which proves that MMR is safe (although the evidence to say it is dangerous was exceptionally poor). There is, for example, no randomised controlled trial. We are presented instead with a huge morass of data, from a number of different studies, all of which are flawed in their own idiosyncratic ways for reasons of cost, competence and so on. A common problem with applying old data to new questions is that these papers and datasets might have a lot of useful information, which was collected very competently to answer the questions which the researchers were interested in at the time, but which isn't perfect for your needs. It's just, perhaps, pretty good.

Smeeth *et al.*, for example, did something called a 'case-control' study, using the GP Research Database. This is a common type of study, where you take a bunch of people with the condition you're looking at ('autism'), and a bunch of people without it, then look to see if there is any difference in how much each group was exposed to the thing you think might be causing the condition ('MMR').

If you care who paid for the study – and I hope you've become a bit more sophisticated than that by now – it was funded by the Medical Research Council. They found around 1,300 people with autism, and then got some 'controls': random people who did not have autism, but with the same age, sex, and general practice. Then they looked to see if vaccination was any more common in the people with autism, or the controls, and found no difference between the two groups. The same researchers also did a systematic review of similar studies in the United States and Scandinavia, and again, pooling the data, found no link between MMR and autism.

There is a practical problem with this kind of research, of course, which I would hope you might spot: most people *do* get the MMR vaccine, so the individuals you're measuring who *didn't* get the vaccine might be unusual in other ways – perhaps their parents have refused the vaccine for ideological or cultural reasons, or the child has a pre-existing physical health problem – and those factors might themselves be related to autism. There's little you can do in terms of study design about this potential 'confounding variable', because as we said, you're not likely to do a randomised controlled trial in which you randomly don't give children vaccines: you just throw the result into the pot with the rest of the information, in order to reach your verdict. As it happens, Smeeth *et al.* went to great lengths to make sure their controls were representative. If you like, you can read the paper and decide if you agree.

So 'Smeeth' was a 'case-control study', where you compare groups which had the outcome or not, and look at how common the exposure was in each group. In Denmark, Madsen *et al.* did the opposite kind of study, called a 'cohort study': you compare groups which had the exposure or not, in order to see whether there is any variation in the outcome. In this specific case, then, you take two groups of people, who either had MMR or didn't, and then check later to see if the rate of autism is any different between the two groups.

This study was big – very big – and included all the children born in Denmark between January 1991 and December 1998. In Denmark there is a system of unique personal identification numbers, linked to vaccination registers and information about the diagnosis of autism, which made it possible to chase up almost all the children in the study. This was a pretty impressive achievement, since there were 440,655 children who were vaccinated, and 96,648 who were unvaccinated. No difference was found between vaccinated and unvaccinated children, in the rates of autism or autistic spectrum disorders, and no association between development of autism and age at vaccination.

Anti-MMR campaigners have responded to this work by saying that only a small number of children are harmed by the vaccine, which seems to be inconsistent with their claims that MMR is responsible for a massive upswing in diagnoses of autism. In any case, if a vaccine caused an adverse reaction in a very small number of people, that would be no surprise – it would be no different from any other medical intervention (or, arguably, any human activity), and there would be, surely, no story.

As with all studies, there are problems with this huge study. The follow-up of diagnostic records ends one year (31 December 1999) after the last day of admission to the cohort: so, because autism comes on after the age of one year, the children born later in the cohort would be unlikely to have shown up with autism by the end of the follow-up period. But this is flagged up in the study, and you can decide whether you think it undermines its overall findings. I don't think it's much of a problem. That's my verdict, and I think you might agree that it's not a particularly foolish one. It did run from January 1991, after all.

This is the kind of evidence you will find in the Cochrane review, which found, very simply, that 'existing evidence on the safety and effectiveness of MMR vaccine supports current policies of mass immunisation aimed at global measles eradication in order to reduce morbidity and mortality associated with mumps and rubella'.

It also contained multiple criticisms of the evidence it reviewed, which, bizarrely, has been seized upon by various commentators to claim that there was some kind of stitch-up. The review was heading towards a conclusion that MMR was risky, they say, if you read the content, but then, out of nowhere, it produced a reassuring conclusion, doubtless because of hidden political pressure.

The *Daily Mail*'s Melanie Phillips, a leading light of the anti-vaccination movement, was outraged by what she thought she had found: 'It said that no fewer than nine of the most celebrated studies that have been used against [Andrew Wakefield] were unreliable in the way they were constructed.' Of course it did. I'm amazed it wasn't more. Cochrane reviews are *intended* to criticise papers.

Scientific 'evidence' in the media

But the newspapers in 2002 had more than just worried parents. There was a smattering of science to keep things going: you will remember computer-generated imagery of viruses and gut walls, perhaps, and stories about laboratory findings. Why have I not mentioned those?

For one thing, these important scientific findings were being reported in newspapers and magazines, and at meetings, in fact anywhere except proper academic journals where they could be read and carefully appraised. In May, for example, Wakefield 'exclusively revealed' that 'more than 95 per cent of those who had the virus in their gut had MMR as their only documented exposure to measles'. He doesn't appear to have revealed this in a peer-reviewed academic journal, but in a weekend colour supplement.

Other people started popping up all over the place, claiming to have made some great finding, but never publishing their research in proper, peer-reviewed academic journals. A pharmacist in Sunderland called Dr Paul Shattock was reported on the *Today* programme, and in several national newspapers, to have identified a distinct subgroup of children with autism resulting from MMR. Dr Shattock is very active on anti-immunisation websites, but he still doesn't seem to have got round to publishing this important work years later, even though the Medical Research Council suggested in 2002 that he should 'publish his research and come forward to the MRC with positive proposals'.

Meanwhile Dr Arthur Krigsman, paediatric gastrointestinal consultant at the New York School of Medicine, was telling hearings in Washington that he had made all kinds of interesting findings in the bowels of autistic children, using endoscopes. This was lavishly reported in the media. Here is the *Daily Telegraph*:

> Scientists in America have reported the first independent corroboration of the research findings of Dr Andrew Wakefield. Dr Krigsman's discovery is significant because it independently supports Dr Wakefield's conclusion that a previously unidentified and devastating combination of bowel and brain disease is afflicting young children – a claim that the Department of Health has dismissed as 'bad science'.

To the best of my knowledge – and I'm pretty good at searching for this stuff – Krigsman's new scientific research findings which corroborate Andrew Wakefield's have never been published in an academic journal: certainly there is no trace of them on Pubmed, the index of nearly all medical academic articles.

In case the reason why this is important has not sunk in, let me explain again. If you visit the premises of the Royal Society in London, you'll see its motto proudly on display: '*Nullius in verba*' – 'On the word of no one'. What I like to imagine this refers to, in my geeky way, is the importance of publishing proper scientific papers if you want people to pay attention to your work. Dr Arthur Krigsman has been claiming for years now that he has found evidence linking MMR to autism and bowel disease. Since he hasn't published his findings, he can claim them until he's blue in the face, because until we can see exactly what he did, we can't see what flaws there may be in his methods. Maybe he didn't select the subjects properly. Maybe he measured the wrong things. If he doesn't write it up formally, we can never know, because that is what scientists do: write papers, and pull them apart to see if their findings are robust.

Krigsman and others' failures to publish in peer-reviewed academic journals weren't isolated incidents. In fact it's still going on, years later. In 2006, exactly the same thing was happening again. 'US Scientists Back Autism Link to MMR', squealed the *Telegraph*. 'Scientists Fear MMR Link to Autism', roared the *Mail*. 'US Study Supports Claims of MMR Link to Autism', croaked *The Times* a day later.

What was this frightening new data? These scare stories were based on a poster presentation, at a conference yet to occur, on research not yet completed, by a man with a track record of announcing research that never subsequently appears in an academic journal. In fact, astonishingly, four years later, it was Dr Arthur Krigsman again. The story this time was different: he had found genetic material (RNA) from vaccine-strain measles virus in some gut samples from children with autism and bowel problems. If true, this would fit with Wakefield's theory, which by 2006 was lying in tatters. We might also mention that Wakefield and Krigsman are doctors together at Thoughtful House, a private autism clinic in the USA offering eccentric treatments for developmental disorders.

The *Telegraph* went on to explain that Krigsman's most recent unpublished claim was replicating similar work from 1998 by Dr Andrew Wakefield, and from 2002 by Professor John O'Leary. This was, to say the least, a mis-statement. There is no work from 1998 by Wakefield which fits the *Telegraph*'s claim – at least not in PubMed that I can find. I suspect the newspaper was confused about the infamous *Lancet* paper on MMR, which by 2004 had already been partially retracted.

There are, however, two papers suggesting that traces of genetic material from the measles virus have been found in children. They have received a mountain of media coverage over half a decade, and yet the media have remained studiously silent on the published evidence suggesting that they were false positives, as we will now see.

One is from Kawashima *et al.* in 2002, also featuring Wakefield as an author, in which it is claimed that genetic material from measles vaccine was found in blood cells. Doubt is cast on this both by attempts to replicate it, showing where the false positives probably appeared, and by the testimony of Nick Chadwick, the PhD student whose work we described above. Even Andrew Wakefield himself no longer relies on this paper.

The other is O'Leary's paper from 2002, also featuring Wakefield as an author, which produced evidence of measles RNA in tissue samples from children. Futher experiments, again, have illustrated where the false positives seem to have arisen, and in 2004, when Professor Stephen Bustin was examining the evidence for the legal aid case, he explained how he established to his satisfaction – during a visit to the O'Leary lab – that these were false positives due to contamination and inadequate experimental methods. He has shown, firstly, that there were no 'controls' to check for false positives (contamination is a huge risk when you are looking for minuscule traces of genetic material, so you generally run 'blank' samples to make sure they do come out blank); he found calibration problems with the machines; problems with log books; and worse. He expanded on this at enormous length in a US court case on autism and vaccines in 2006. You can read his detailed explanation in full online. To my astonishment not one journalist in the UK has ever bothered to report it.

Both of these papers claiming to show a link received blanket media coverage at the time, as did Krigsman's claims.

What they didn't tell you

In the May 2006 issue of the *Journal of Medical Virology* there was a very similar study to the one described by Krigsman, only this one had actually been published, by Afzal *et al*. It looked for measles RNA in children with regressive autism after MMR vaccination, much like the unpublished Krigsman study, and it used tools so powerful they could detect measles RNA down to single-figure copy numbers. It found no evidence of the magic vaccine-strain measles RNA to implicate MMR. Perhaps because of that unfrightening result, the study was loudly ignored by the press.

Because it has been published in full, I can read it, and pick holes in it, and I am more than happy to do so: because science is about critiquing openly published data and methodologies, rather than press-released chimeras, and in the real world all studies have some flaws, to a greater or lesser extent. Often they are practical ones: here, for example, the researchers couldn't get hold of the tissue they ideally would have used, because they could not get ethics committee approval for intrusive procedures like lumbar punctures and gut biopsies on children (Wakefield did manage to obtain such samples, but he is, we should remember, currently going through a GMC professional conduct hearing over the issue).

Surely they could have borrowed some existing samples, from children said to be damaged by vaccines? You'd have thought so. They report in the paper that they tried to ask anti-MMR researchers – if that's not an unfair term – whether they could borrow some of their tissue samples to work on. They were ignored.*

Afzal *et al.* was not reported in the media, anywhere at all, except by me, in my column.

This is not an isolated case. Another major paper was published in the leading academic journal *Pediatrics* a few months later – to complete media silence – again suggesting very strongly that the earlier results from Kawashima and O'Leary were in error, and false positives. D'Souza *et al.* replicated the earlier experiments very closely, and in some respects more carefully: most import-

---

* 'The groups of investigators that either had access to original autism specimens or investigated them later for measles virus detection were invited to take part in the study but failed to respond. Similarly, it was not possible to obtain clinical specimens of autism cases from these investigators for independent investigations.'

antly, it traced out the possible routes by which a false positive could have occurred, and made some astonishing findings.

False positives are common in PCR, because it works by using enzymes to replicate RNA, so you start with a small amount in your sample, which is then 'amplified up', copied over and over again, until you have enough to measure and work with. Beginning with a single molecule of genetic material, PCR can generate 100 billion similar molecules in an afternoon. Because of this, the PCR process is exquisitely sensitive to contamination – as numerous innocent people languishing in jail could tell you – so you have to be very careful, and clean up as you go.

As well as raising concerns about contamination, D'Souza also found that the O'Leary method might have accidentally amplified the wrong bits of RNA.

Let's be clear: this is absolutely not about criticising individual researchers. Techniques move on, results are sometimes not replicable, and not all double-checking is practical (although Bustin's testimony is that standards in the O'Leary lab were problematic). But what is striking is that the media rabidly picked up on the original frightening data, and then completely ignored the new reassuring data. This study by D'Souza, like Afzal before it, was unanimously ignored by the media. It was covered, by my count, in: my column; one Reuters piece which was picked up by nobody; and one post on the lead researcher's boyfriend's blog (where he talked about how proud he was of his girlfriend). Nowhere else.*

You could say, very reasonably, that this is all very much par for the course: newspapers report the news, and it's not very

---

* In 2008, just as this chapter was being put to bed, some journalists deigned – miraculously – to cover a PCR experiment with a negative finding. It was misreported as the definitive refutation of the entire MMR–autism hypothesis. This was a childish overstatement, and that doesn't help anyone either. I am not hard to please.

interesting if a piece of research comes out saying something is safe. But I would argue – perhaps sanctimoniously – that the media have a special responsibility in this case, because they themselves demanded 'more research', and moreover because *at the very same time* that they were ignoring properly conducted and fully published negative findings, they were talking up scary findings from an unpublished study by Krigsman, a man with a track record of making scary claims which remain unpublished.

MMR is not an isolated case in this regard. You might remember the scare stories about mercury fillings from the past two decades: they come around every few years, usually accompanied by a personal anecdote in which fatigue, dizziness and headaches are all vanquished following the removal of the fillings by one visionary dentist. Traditionally these stories conclude with a suggestion that the dental establishment may well be covering up the truth about mercury, and a demand for more research into its safety.

The first large-scale randomised control trials on the safety of mercury fillings were published recently, and if you were waiting to see these hotly anticipated results, personally demanded by journalists on innumerable newspapers, you'd be out of luck, because they were reported nowhere. Nowhere. A study of more than 1,000 children, where some were given mercury fillings and some mercury-free fillings, measuring kidney function and neurodevelopmental outcomes like memory, coordination, nerve conduction, IQ and so on over several years. It was a well-conducted study. There were no significant differences between the two groups. That's worth knowing about if you've ever been scared by the media's reports on mercury fillings – and by God, you'd have been scared.

Panorama featured a particularly chilling documentary in 1994 called *The Poison in Your Mouth*. It opened with dramatic footage of men in full protective gear rolling barrels of mercury around. I'm not giving you the definitive last word on mercury here. But I think we can safely assume there is no *Panorama* documentary in the pipeline covering the startling new research data suggesting that mercury fillings may not be harmful after all.

In some respects this is just one more illustration of how unreliable intuition can be in assessing risks like those presented with a vaccine: not only is it a flawed strategy for this kind of numerical assessment, on outcomes which are too rare for one person to collect meaningful data on them in their personal journey through life; but the information you are fed by the media about the wider population is ludicrously, outrageously, criminally crooked. So at the end of all this, what has the British news media establishment achieved?

Old diseases return

It's hardly surprising that the MMR vaccination rate has fallen from 92 per cent in 1996 to 73 per cent today. In some parts of London it's down to 60 per cent, and figures from 2004–05 showed that in Westminster only 38 per cent per cent of children had both jabs by the age of five.*

It is difficult to imagine what could be driving this, if not a brilliantly successful and well-coordinated media anti-MMR campaign, which pitched emotion and hysteria against scientific evidence. Because people listen to journalists: this has been demonstrated repeatedly, and not just with the kinds of stories in this book.

A 2005 study in the *Medical Journal of Australia* looked at mammogram bookings, and found that during the peak media coverage of Kylie Minogue's breast cancer, bookings rose by 40 per cent. The increase among previously unscreened women in the forty-to-sixty-nine-year age group was 101 per cent. These surges were unprecedented. And I'm not cherry-picking: a

---

* Not 11.7 per cent as claimed in the *Telegraph* and the *Daily Mail* in February and June 2006.

systematic review from the Cochrane Collaboration found five studies looking at the use of specific health interventions before and after media coverage of specific stories, and each found that favourable publicity was associated with greater use, and unfavourable coverage with lower use.

It's not just the public: medical practice is influenced by the media too, and so are academics. A mischievous paper from the *New England Journal of Medicine* in 1991 showed that if a study was covered by the *New York Times*, it was significantly more likely to be cited by other academic papers. Having come this far, you are probably unpicking this study already. Was coverage in the *New York Times* just a surrogate marker for the importance of the research? History provided the researchers with a control group to compare their results against: for three months, large parts of the paper went on strike, and while the journalists did produce an 'edition of record', this newspaper was never actually printed. They wrote stories about academic research, using the same criteria to judge importance that they always had, but the research they wrote about in articles which never saw the light of day saw no increase in citations.

People read newspapers. Despite everything we think we know, their contents seep in, we believe them to be true, and we act upon them, which makes it all the more tragic that their contents are so routinely flawed. Am I extrapolating unfairly from the extreme examples in this book? Perhaps not. In 2008 Gary Schwitzer, an ex-journalist who now works on quantitative studies of the media, published an analysis of five hundred health articles covering treatments from mainstream newspapers in the US. Only 35 per cent of stories were rated satisfactory for whether the journalist had 'discussed the study methodology and the quality of the evidence' (because in the media, as we have seen repeatedly in this book, science is about absolute truth statements from arbitrary authority figures in white coats, rather than clear descriptions of studies, and the reasons why people draw conclusions from them). Only 28 per cent adequately covered benefits, and only 33 per cent adequately covered harms. Articles routinely failed to give any useful quantitative information in absolute terms, preferring unhelpful eye-catchers like '50 per cent higher' instead.

In fact there have been systematic quantitative surveys of the accuracy of health coverage in Canada, Australia and America – I'm trying to get one off the ground in the UK – and the results have been universally unimpressive. It seems to me that the state of health coverage in the UK could well be a serious public health issue.

Meanwhile, the incidence of two of the three diseases covered by MMR is now increasing very impressively. We have the highest number of measles cases in England and Wales since current surveillance methods began in 1995, with cases occurring mostly in children who had not been adequately vaccinated: 971 confirmed cases were reported in 2007 (mostly associated with prolonged outbreaks in travelling and religious communities, where vaccine uptake has been historically low), after 740 cases in 2006 (and the first death since 1992). Seventy-three per cent of cases were in the South-East, and most of those were in London.

Mumps began rising again in 1999, after many years of cases in only double figures: by 2005 the United Kingdom had a mumps epidemic, with around 5,000 notifications in January alone.

A lot of people who campaign against vaccines like to pretend that they don't do much good, and that the diseases they protect against were never very serious anyway. I don't want to force anyone to have their child vaccinated, but equally I don't think anyone is helped by misleading information. By contrast with the unlikely event of autism being associated with MMR, the risks from measles, though small, are real and quantifiable. The Peckham Report on immunisation policy, published shortly after the introduction of the MMR vaccine, surveyed the recent experience of measles in Western countries and estimated that for every 1,000 cases notified, there would be 0.2 deaths, ten hospital admissions, ten neurological complications and forty respiratory complications. These estimates have been borne out in recent minor epidemics in the Netherlands (1999: 2,300 cases in a community philosophically opposed to vaccination, three deaths), Ireland (2000: 1,200 cases, three deaths) and Italy (2002: three deaths). It's worth noting that plenty of these deaths were in previously healthy children, in developed countries, with good healthcare systems.

Though mumps is rarely fatal, it's an unpleasant disease with unpleasant complications (including meningitis, pancreatitis and sterility). Congenital rubella syndrome has become increasingly rare since the introduction of MMR, but causes profound disabilities including deafness, autism, blindness and mental handicap, resulting from damage to the foetus during early pregnancy.

The other thing you will hear a lot is that vaccines don't make much difference anyway, because all the advances in health and life expectancy have been due to improvements in public health for a wide range of other reasons. As someone with a particular interest in epidemiology and public health, I find this suggestion flattering; and there is absolutely no doubt that deaths from measles began to fall over the whole of the past century for all kinds of reasons, many of them social and political as well as medical: better nutrition, better access to good medical care, antibiotics, less crowded living conditions, improved sanitation, and so on.

Life expectancy in general has soared over the past century, and it's easy to forget just how phenomenal this change has been. In 1901, males born in the UK could expect to live to forty-five, and females to forty-nine. By 2004, life expectancy at birth had risen to seventy-seven for men, and eighty-one for women (although of course much of the change is due to reductions in infant mortality).

Life expectancy at birth in the UK

So we are living longer, and vaccines are clearly not the only reason why. No single change is the reason why. Measles incidence dropped hugely over the preceding century, but you would have to work fairly hard to persuade yourself that vaccines had no impact on that. Here, for example, is a graph showing the reported incidence of measles from 1950 to 2000 in the United States.

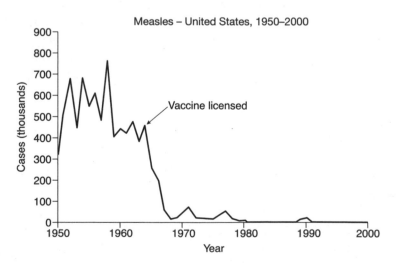

Measles – United States, 1950–2000

The Media's MMR Hoax

For those who think that single vaccines for the components of MMR are a good idea, you'll notice that these have been around since the 1970s, but that a concerted programme of vaccination – and the concerted programme of giving all three vaccinations in one go as MMR – is fairly clearly associated in time with a further (and actually rather definitive) drop in the rate of measles cases.

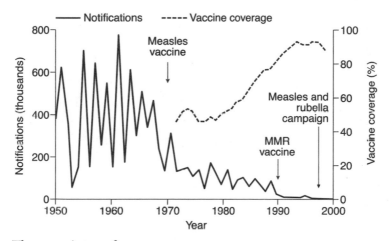

The same is true for mumps:

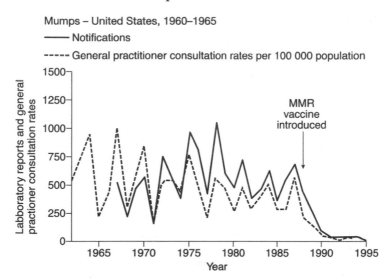

Mumps – United States, 1960–1965

While we're thinking about mumps, let's not forget our epidemic in 2005, a resurgence of a disease many young doctors would struggle even to recognise. Here is a graph of mumps cases from the *BMJ* article that analysed the outbreak:

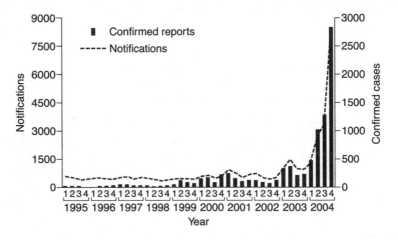

Almost all confirmed cases during this outbreak were in people aged fifteen to twenty-four, and only 3.3 per cent had received the full two doses of MMR vaccine. Why did it affect these people? Because of a global vaccine shortage in the early 1990s.

Mumps is not a harmless disease. I've no desire to scare anyone – and as I said, your beliefs and decisions about vaccines are your business; I'm only interested in how you came to be so incredibly misled – but before the introduction of MMR, mumps was the commonest cause of viral meningitis, and one of the leading causes of hearing loss in children. Lumbar puncture studies show that around half of all mumps infections involve the central nervous system. Mumps orchitis is common, exquisitely painful, and occurs in 20 per cent of adult men with mumps: around half will experience testicular atrophy, normally in only one testicle, but 15 to 30 per cent of patients with mumps orchitis will have it in both testicles, and of these, 13 per cent will have reduced fertility.

I'm not just spelling this out for the benefit of the lay reader: by the time of the outbreak in 2005, young doctors needed to be reminded of the symptoms and signs of mumps, because it had been such an uncommon disease during their training and clinical experience. People had forgotten what these diseases looked like, and in that regard vaccines are a victim of their own success – as we saw in our earlier quote from *Scientific American* in 1888, five generations ago (see page 276).

Whenever we take a child to be vaccinated, we're aware that we are striking a balance between benefit and harm, as with any medical intervention. I don't think vaccination is all that important: even if mumps orchitis, infertility, deafness, death and the rest are no fun, the sky wouldn't fall in without MMR. But taken on their own, lots of other individual risk factors aren't very important either, and that's no reason to abandon all hope of trying to do something simple, sensible and proportionate about them, gradually increasing the health of the nation, along with all the other stuff you can do to the same end.

It's also a question of consistency. At the risk of initiating mass panic, I feel duty bound to point out that if MMR still scares you, then so should everything in medicine, and indeed many of the everyday lifestyle risk exposures you encounter: because there are a huge number of things which are far less well researched, with a far lower level of certainty about their safety. The question would still remain of why you were so focused on MMR. If you wanted to do something constructive about this problem, instead of running a single-issue campaign about MMR you might, perhaps, use your energies more usefully. You could start a campaign for constant automated vigilance of the entirety of the NHS health records dataset for any adverse outcomes associated with any intervention, for example, and I'd be tempted to join you on the barricades.

But in many respects this isn't about risk management, or vigilance: it's about culture, human stories, and everyday human harms. Just as autism is a peculiarly fascinating condition to journalists, and indeed to all of us, vaccination is similarly inviting as a focus for our concerns: it's a universal programme, in conflict with modern ideas of 'individualised care'; it's bound up with government; it involves needles going into children; and it offers the opportunity to blame someone, or something, for a dreadful tragedy.

Just as the causes of these scares have been more emotional than anything else, so too has much of the harm. Parents of children with autism have been racked with guilt, doubt and endless self-recrimination over the thought that they themselves are responsible for inflicting harm upon their own child. This distress has been demonstrated in countless studies: but so close to the end, I don't want to introduce any more research papers.

There is one quote that I find – although she would perhaps complain about my using it – both moving and upsetting. It's from Karen Prosser, who featured with her autistic son Ryan in the Andrew Wakefield video news release from the Royal Free Hospital in 1998. 'Any mother who has a child wants it to be normal,' she says. 'To then find out your child might be genetically autistic is tragic. To find out that it was caused by a vaccine, that you agreed to have done … is just devastating.'

AND ANOTHER THING

I could go on. As I write this in May 2008, the media are still pushing a celebrity-endorsed 'miracle cure' (and I quote) for dyslexia, invented by a millionaire paint entrepreneur, despite the abysmal evidence to support it, and despite customers being at risk of simply losing their money anyway, because the company seems to be going into administration; the newspapers are filled with an amazing story about a finger that 'grew back' through the use of special sciencey 'pixie dust' (I quote again), although the claim has been around for three years, unpublished in any academic journal, and severed fingertips grow back by themselves anyway; more 'hidden data' scandals are exposed from the vaults of big pharma every month; quacks and cranks continue to parade themselves on television quoting fantastical studies to universal approbation; and there will always be new scares, because they sell so very well, and they make journalists feel alive.

To anyone who feels their ideas have been challenged by this book, or who has been made angry by it – to the people who feature in it, I suppose – I would say this: You win. You really do. I would hope there might be room for you to reconsider, to change your stance in the light of what might be new information (as I will happily do, if there is ever an opportunity to update this book). But you will not need to, because, as we both know, you collectively have almost full-spectrum dominance: your own slots in every newspaper and magazine in Britain, and front-page coverage for your scare stories. You affect outsider swagger, bizarrely, from the sofas of daytime television. Your ideas – bogus though they may be – have immense superficial plausibility, they can be expressed rapidly, they are endlessly repeated, and they are believed by enough people for you to make very comfortable livings, and to have enormous cultural influence. You win.

It's not the spectacular individual stories that are the problem, so much as the constant daily grind of stupid little ones. This will not end, and so I will now abuse my position by telling you, very briefly, exactly what I think is wrong, and some of what can be done to fix it.

The process of obtaining and interpreting evidence isn't taught in schools, nor are the basics of evidence-based medicine and epidemiology, yet these are obviously the scientific issues which are most on people's minds. This is not idle speculation. You will remember that this book began by noticing that there has never been an exhibit on evidence-based medicine in London's Science Museum.

A five-decade survey of post-war science coverage in the UK by the same institution shows – and this is officially the last piece of data in the book – that in the 1950s science reporting was about engineering and inventions, but by the 1990s everything had changed. Science coverage now tends to come from the world of medicine, and the stories are of what will kill you, or save you. Perhaps it is narcissism, or fear, but the science of health is important to people, and at the very time when we need it the most, our ability to think around the issue is being energetically distorted by the media, corporate lobbies and, frankly, cranks.

Without anybody noticing, bullshit has become an extremely important public health issue, and for reasons that go far beyond the obvious hysteria around immediate harms: the odd measles tragedy, or a homeopath's unnecessary malaria case. Doctors today are keen – as it said in our medical school notes

– to work 'collaboratively with the patient towards an optimum health outcome'. They discuss evidence with their patients, so that they can make their own decisions about treatments.

I don't generally talk or write about being a doctor – it's mawkish and tedious, and I've no desire to preach from author-ity – but working in the NHS you meet patients from every conceivable walk of life, in huge numbers, discussing some of the most important issues in their lives. This has consistently taught me one thing: people aren't stupid. Anybody can under-stand anything, as long as it is clearly explained – but more than that, if they are sufficiently interested. What determines an audience's understanding is not so much scientific knowledge, but motivation: patients who are ill, with an important decision to make about treatment, can be very motivated indeed.

But journalists and miracle-cure merchants sabotage this process of shared decision-making, diligently, brick by brick, making lengthy and bogus criticisms of the process of system-atic review (because they don't like the findings of just one), extrapolating from lab-dish data, misrepresenting the sense and value of trials, carefully and collectively undermining the nation's understanding of the very notion of what it means for there to be evidence for an activity. In this regard they are, to my mind, guilty of an unforgivable crime.

You'll notice, I hope, that I'm more interested in the cultural impact of nonsense – the medicalisation of everyday life, the undermining of sense – and in general I blame systems more than particular people. While I do go through the background of some individuals, this is largely to illustrate the extent to which they have been misrepresented by the media, who are so desperate to present their favoured authority figures as somehow mainstream. I am not surprised that there are indi-vidual entrepreneurs, but I am unimpressed that the media carry their assertions as true. I am not surprised that there are people with odd ideas about medicine, or that they sell those

ideas. But I am spectacularly, supremely, incandescently unimpressed when a university starts to offer BSc science courses in
them. I do not blame individual journalists (for the most part),
but I do blame whole systems of editors, and the people who
buy newspapers with values they profess to despise. Specifically,
I do not blame Andrew Wakefield for the MMR scare (although
he's done things I hope I would not), and I find it – let's be very
clear once again – spectacularly distasteful that the media are
now revving up to hold him singly responsible for their own
crimes, in regard to that débâcle.

Similarly, while I could reel out a few stories of alternative
therapists' customers who've died unnecessarily, it seems to me
that people who choose to see alternative therapists (except for
nutrition therapists, who have worked *very* hard to confuse the
public and to brand themselves as conventional evidence-based
practitioners) make that choice with their eyes open, or at least
only half closed. To me this is not a situation of businessmen
exploiting the vulnerable, but is rather, as I seem to keep saying,
a bit more complicated than that. We love this stuff, and we love
it for some fascinating reasons, which we could ideally spend a
lot more time thinking and talking about.

Economists and doctors talk about 'opportunity costs', the
things you could have done, but didn't, because you were
distracted by doing something less useful. To my mind, the
greatest harm posed by the avalanche of nonsense we have seen
in this book is best conceived of as the 'opportunity cost of bullshit'.

We have have somehow become collectively obsessed with
these absurd, thinly evidenced individual tinkerings in diet,
distracting us from simple healthy eating advice; but more than
that, as we saw, distracting us from the other important lifestyle
risk factors for ill health which cannot be sold, or commodified.

Doctors, similarly, have been captivated by the commercial
success of alternative therapists. They could learn from the best

of the research into the placebo effect, and the meaning response in healing, and apply that to everyday clinical practice, augmenting treatments which are in themselves also effective: but instead, there is a fashion among huge numbers of them to indulge childish fantasies about magic pills, massages or needles. That is not forward-looking, or inclusive, and it does nothing about the untherapeutic nature of rushed consultations in decaying buildings. It also requires, frequently, that you lie to patients. 'The true cost of something,' as the *Economist* says, 'is what you give up to get it.'

On a larger scale, many people are angry about the evils of the pharmaceutical industry, and nervous about the role of profit in healthcare; but these are formless and uncalibrated intuitions, so the valuable political energy that comes from this outrage is funnelled – wasted – through infantile issues like the miraculous properties of vitamin pills, or the evils of MMR. Just because big pharma can behave badly, that does not mean that sugar pills work better than placebo, nor does it mean that MMR causes autism. Whatever the wealthy pill peddlers try to tell you, with their brand-building conspiracy theories, big pharma isn't *afraid* of the food supplement pill industry, it *is* the food supplement pill industry. Similarly, big pharma isn't frightened for its profits because popular opinion turned against MMR: if they have any sense, these companies are relieved that the public is obsessed with MMR, and is thus distracted from the other far more complex and real issues connected with the pharmaceutical business and its inadequate regulation.

To engage meaningfully in a political process of managing the evils of big pharma, we need to understand a little about the business of evidence: only then can we understand why transparency is so important in pharmaceutical research, for example, or the details of how it can be made to work, or concoct new and imaginative solutions.

But the greatest opportunity cost comes, of course, in the media, which has failed science so spectacularly, getting stuff wrong, and dumbing down. No amount of training will ever improve the wildly inaccurate stories, because newspapers already have specialist health and science correspondents who understand science. Editors will always – cynically – sideline those people, and give stupid stories to generalists, for the simple reason that they want stupid stories. Science is beyond their intellectual horizon, so they assume you can just make it up anyway. In an era when mainstream media is in fear for its life, their claims to act as effective gatekeepers to information are somewhat undermined by the content of pretty much every column or blog entry I've ever written.

To academics, and scientists of all shades, I would say this: you cannot ever possibly prevent newspapers from printing nonsense, but you can add your own sense into the mix. Email the features desk, ring the health desk (you can find the switchboard number on the letters page of any newspaper), and offer them a piece on something interesting from your field. They'll turn you down. Try again. You can also toe the line by not writing stupid press releases (there are extensive guidelines for communicating with the media online), by being clear about what's speculation in your discussions, by presenting risk data as 'natural frequencies', and so on. If you feel your work – or even your field – has been misrepresented, then complain: write to the editor, the journalist, the letters page, the readers' editor, the PCC; put out a press release explaining why the story was stupid, get your press office to harrass the paper or TV station, use your title (it's embarrassing how easy they are to impress), and offer to write them something yourself.

The greatest problem of all is dumbing down. Everything in the media is robbed of any scientific meat, in a desperate bid to seduce an imaginary mass who aren't interested. And why should they be? Meanwhile the nerds, the people who studied biochemistry but who now work in middle management at Woolworths, are neglected, unstimulated, abandoned. There are intelligent people out there who want to be pushed, to keep their knowledge and passion for science alive, and neglecting them comes at a serious cost to society. Institutions have failed in this regard. The indulgent and well-financed 'public engagement with science' community has been worse than useless, because it too is obsessed with taking the message to everyone, rarely offering stimulating content to the people who are already interested.

Now you don't need these people. Start a blog. Not everyone will care, but some will, and they will find your work. Unmediated access to niche expertise is the future, and you know, science isn't hard – academics around the world explain hugely complicated ideas to ignorant eighteen-year-olds every September – it just requires motivation. I give you the CERN podcast, the Science in the City mp3 lecture series, blogs from profs, open access academic journal articles from PLOS, online video archives of popular lectures, the free editions of the Royal Statistical Society's magazine *Significance*, and many more, all out there, waiting for you to join them. There's no money in it, but you knew that when you started on this path. You will do it because you know that knowledge is beautiful, and because if only a hundred people share your passion, that is enough.

FURTHER READING AND ACKNOWLEDGEMENTS

I have done my absolute best to keep these references to a minimum, as this is supposed to be an entertaining book, not a scholarly text. More useful than references, I would hope, are the many extra materials available on www.badscience.net, including recommended reading, videos, a rolling ticker of interesting news stories, updated references, activities for schoolchildren, a discussion forum, everything I've ever written (except this book, of course), advice on activism, links to science communication guidelines for journalists and academics, and much more. I will always try to add to it as time passes.

There are some books which really stand out as genuinely excellent, and I am going to use my last ink to send you their way. Your time will not be wasted on them.

Testing Treatments by Imogen Evans, Hazel Thornton and Iain Chalmers is a book on evidence-based medicine specifically written for a lay audience by two academics and a patient. It is also free to download online from www.jameslindlibrary.org. *How to Read a Paper* by Professor Greenhalgh is the standard medical textbook on critically appraising academic journal articles. It's readable, short, and it would be a best-seller if it wasn't unnecessarily overpriced.

Irrationality by Stuart Sutherland makes a great partner with *How We Know What Isn't So* by Thomas Gilovich, as both cover different aspects of social science and psychology research into irrational behaviour, while *Reckoning with Risk* by Gerd Gigerenzer comes at the same problems from a more mathematical perspective.

Meaning, Medicine and the 'Placebo Effect' by Daniel Moerman is excellent, and you should not be put off by the fact that it is published under an academic imprint.

There are now endless blogs by like-minded people which have sprung from nowhere over the past few years, to my enormous delight, onto my computer screen. They often cover science news better than the mainstream media, and the feeds of some of the most entertaining fellow-travellers are aggregated at the website badscienceblogs.net. I enjoy disagreeing with many of them – viciously – on a great many things.

And lastly, the most important references of all are to the people by whom I have been taught, nudged, reared, influenced, challenged, supervised, contradicted, supported, and most importantly entertained. They are (missing too many, and in very little order): Emily Wilson, Ian Sample, James Randerson, Alok Jha, Mary Byrne, Mike Burke, Ian Katz, Mitzi Angel, Robert Lacey, Chris Elliott, Rachel Buchanan, Alan Rusbridger, Pat Kavanagh, the inspirational badscience bloggers, everyone who has ever sent me a tip about a story on ben@badscience.net, Iain Chalmers, Lorne Denny, Simon Wessely, Caroline Richmond, John Stein, Jim Hopkins, David Colquhoun, Catherine Collins, Matthew Hotopf, John Moriarty, Alex Lomas, Andy Lewis, Trisha Greenhalgh, Gimpy, shpalman, Holfordwatch, Positive Internet, Jon, Liz Parratt, Patrick Matthews, Ian Brown, Mike Jay, Louise Burton, John King, Cicely Marston, Steve Rolles, Hettie, Mark Pilkington, Ginge Tulloch, Matthew Tait, Cathy Flower, my mum, my dad, Reg, Josh, Raph, Allie, and the fabulous Amanda Palmer.

NOTES

Chapter 1: Matter

2 'We sent Alex': *Daily Mirror* (4 January 2003)

6 'The candles work by': http://www.bbc.co.uk/wales/southeast/sites/mind/pages/hopi.shtml

6 'a paper published': Seely DR, Quigley SM, Langman AW. Ear candles – efficacy and safety. Laryngoscope (October1996); 106 (10): 1226–9.

7 'a published study': Ibid.

11 'These cleansing and': Green EC, Honwana A. Indigenous healing of war-affected children in Africa. IK Notes No. 10. Knowledge and Learning Center Africa Region, World Bank Washington (1999), available: http://www.africaaction.org/docs99/viol9907.htm

Chapter 4: Homeopathy

39 'In one study': Marshall T. Reducing unnecessary consultation – a case of NNT? Bandolier (1997); 44 (4): 1–3

41 'But the point of the study': MacManus MP, Matthews JP, Wada M, Wirth A, Worotniuk V, Ball DL. Unexpected long-term survival after low-dose palliative radiotherapy for non-small cell lung cancer. Cancer (1 March 2006); 106 (5): 1110–16.

46 'a very theatrical trial': Majeed AW et al. Randomised, prospective, single-blind comparison of laparoscopic versus small-incision cholecystectomy. Lancet (13 April 1996); 347 (9007): 989–94.

46 'a review of blinding': Schultz KF, Chalmers I, Hayes RJ, Altman DG. Empirical evidence of bias: Dimensions of methodological quality associated with estimates of treatment effects in controlled trials. JAMA (1995); 273: 408–12

46 'review of trials': Ernst E, White AR. Acupuncture for back pain: a meta-analysis of randomised controlled trials. Arch Int Med (1998); 158: 2235–41

47 Diagram: Ibid.

48 'Let us take out': van Helmont JB. *Oriatrike, or Physick Refined: The Common Errors Therein Refuted and the Whole are Reformed and Rectified*. Lodowick-Loyd (1662): 526. Available at (http://www.jameslindlibrary.org

50 'two landmark studies': Khan KS, Daya S, Jadad AR. The importance of quality of primary studies in producing unbiased systematic reviews. Arch Intern Med (1996), 156: 661–6; Moher D, Pham B, Jones A et al. Does quality of reports of randomised trials affect estimates of intervention efficacy reported in meta-analyses? Lancet (1998); 352: 609–13

53 Diagram: Ernst E, Pittler MH. Re-analysis of previous meta-analysis of clinical trials of homeopathy. J Clin Epi (2000); 53 (11): 1188

56 'A landmark meta-analysis': Shang A, Huwiler-Müntener K, Nartey L, Jüni P, Dörig S, Sterne JA, Pewsner D, Egger M. Are the clinical effects of homoeopathy placebo effects? Comparative study of placebo-controlled trials of homoeopathy and allopathy. Lancet (27 August–2 September 2005); 366 (9487): 726–32

58 'One study actually thought': Tallon D, Chard J, Dieppe P. Relation between agendas of the research community and the research consumer. Lancet (2000); 355: 2037–40

61 'They might flick': BBC Radio 4 Case Notes (19 July 2005)

Chapter 5: The Placebo Effect

63 'Shall [the placebo]': The placebo in medicine. Med Press (18 June 1890): 642

64 'Henry Beecher': Beecher HK. The powerful placebo. JAMA (24 December 1955); 159 (17): 1602–6

64 'Peter Parker': Skrabanek P, McCormick J. *Fads and Fallacies in Medicine*. Prometheus Books (1990)

66 'Daniel Moerman': Moerman DE. General medical effectiveness and human biology: placebo effects in the treatment of ulcer disease. Med Anth Quarterly (August 1983); 14; 4: 3–16

67 'in a different dataset': de Craen AJ, Moerman DE, Heisterkamp SH, Tytgat GN, Tijssen JG, Kleijnen J. Placebo effect in the treatment of duodenal ulcer. Br J Clin Pharmacol (December 1999); 48 (6): 853–60

68 'Blackwell [1972]': Blackwell B, Bloomfield SS, Buncher CR. Demonstration to medical students of placebo responses and non-drug factors. Lancet (10 June 1972); 1 (7763): 1279–82

68 'Another study': Schapira K, McClelland HA, Griffiths NR, Newell DJ. Study on the effects of tablet colour in the treatment of anxiety states. BMJ (23 May 1970); 1 (5707): 446–9

68 'a survey of the colour': de Craen AJ, Roos PJ, Leonard de Vries A, Kleijnen J. Effect of colour of drugs: systematic review of perceived effect of drugs and of their effectiveness. BMJ (21–28 December1996); 313 (7072): 1624–6

68 'In 1970': Hussain MZ, Ahad A. Tablet colour in anxiety states. BMJ (22 August 1970); 3 (5720): 466

69 'Route of administration': Grenfell RF, Briggs AH, Holland WC. Double blind study of the treatment of hypertension. JAMA (1961) 176: 124–8; De Craen AJM, Tijssen JGP, de Gans J, Kleijnen J. Placebo effect in the acute treatment of migraine: subcutaneous placebos are better than oral placebos. J Neur (2000) 247: 183–8; Gracely RH, Dubner R, McGrath PA. Narcotic analgesia: fentanyl reduces the intensity but not the unpleasantness of painful tooth pulp sensations. Science (23 March 1979); 203 (4386): 1261–3

69 'the bizarre story of packaging': Kaptchuk TJ, Stason WB, Davis RB, Legedza AR, Schnyer RN, Kerr CE, Stone DA, Nam BH, Kirsch I, Goldman RH. Sham device v inert pill: randomised controlled trial of two placebo treatments. BMJ (18 February 2006); 332 (7538): 391–7

69 'Branthwaite and Cooper': Branthwaite A, Cooper P. Analgesic effects of branding in treatment of headaches. BMJ (Clin Res ed) (1981); 282: 1576–8

70 'a recent study': Waber et al. Commercial features of placebo and therapeutic efficacy. JAMA (2008); 299: 1016–17

70 'a paper currently in press': Ginoa F. Do we listen to advice just because we paid for it? The impact of advice cost on its use. Organizational Behavior and Human Decision Processes 2008,

Article in Press (published online 25 April 2008).
http://dx.doi.org/10.1016/j.obhdp.2008.03.001

70 'Montgomery and Kirsch': Montgomery GH, Kirsch I.
Mechanisms of placebo pain reduction: an empirical
investigation. Psych Science (1996) 7: 174–6

71 'a placebo-controlled trial': Cobb LA, Thomas GI, Dillard DH,
Merendino KA, Bruce RA. An evaluation of internal-
mammary-artery ligation by a double-blind technic. N Eng J
Med (28 May 1959); 260 (22): 1115–18

72 'A Swedish study': Linde C, Gadler F, Kappenberger L, Rydén L.
Placebo effect of pacemaker implantation in obstructive
hypertrophic cardiomyopathy. PIC Study Group. Am J Cardiol
(15 March 1999); 83 (6): 903–7

72 'Electrical machines have': Johnson AG. Surgery as a placebo.
Lancet (22 October 1994); 344 (8930): 1140–2

72 'an elegant study': Crum AJ, Langer EJ. Mind-set matters:
exercise and the placebo effect. Psych Science (February 2007);
18 (2): 165–71

73 'Gryll and Katahn': Gryll SL, Katahn M. Situational factors
contributing to the placebos effect. Psychopharmacology (Berl)
(1978); 57: 253–61

74 'Gracely [1985]': Gracely RH, Dubner R, Deeter WR, Wolskee
PJ. Clinicians' expectations influence placebo analgesia. Lancet
(5 January 1985); 1 (8419): 43

75 'In 1987, Thomas': Thomas KB. General practice consultations:
is there any point in being positive? BMJ (Clin Res ed) (9 May
1987); 294 (6581): 1200–2

76 'Raymond Tallis': Tallis R. *Hippocratic Oaths: Medicine and its
Discontents*. Atlantic (2004)

77 'Quesalid': Lévi-Strauss C. The sorcerer and his magic. In
Structural Anthropology (trans Jacobson C, Schoef BG). Basic
Books (1963)

78 'a classic study from 1965': Park LC, Covi L. Nonblind placebo
trial: an exploration of neurotic patients' responses to placebo
when its inert content is disclosed. Arch Gen Psych (April
1965); 12: 36–45

78 'Dr Stewart Wolf': Wolf S. Effects of suggestion and
conditioning on the action of chemical agents in human

subjects; the pharmacology of placebos. J Clin Invest (January 1950); 29 (1): 100–9

79 'It's been shown': de la Fuente-Fernández R, Ruth TJ, Sossi V, Schulzer M, Calne DB, Stoessl AJ. Expectation and dopamine release: mechanism of the placebo effect in Parkinson's disease. Science (10 August 2001); 293 (5532): 1164–6

79 'Zubieta [2005]': Zubieta JK, Bueller JA, Jackson LR, Scott DJ, Xu Y, Koeppe RA, Nichols TE, Stohler CS. Placebo effects mediated by endogenous opioid activity on mu-opioid receptors. J Neur (24 August 2005); 25 (34): 7754–62

80 'Researchers have measured': Ader R, Cohen N. Behaviorally conditioned immunosuppression. Psychosom Med (July–August 1975); 37 (4): 333–40

80 'researchers gave healthy': Goebel MU, Trebst AE, Steiner J, Xie YF, Exton MS, Frede S, Canbay AE, Michel MC, Heemann U, Schedlowski M. Behavioral conditioning of immunosuppression is possible in humans. FASEB J (December 2002); 16 (14): 1869–73

80 'Researchers have even': Buske-Kirschbaum A, Kirschbaum C, Stierle H, Lehnert H, Hellhammer D. Conditioned increase of natural killer cell activity (NKCA) in humans. Psychosom Med (March–April 1992); 54 (2): 123–32

80 'From survey data': Goodwin JS, Goodwin JM, Vogel AV. Knowledge and use of placebos by house officers and nurses. Ann Intern Med (July 1979); 91 (1): 106–10

80 'You are a placebo responder': ?Meaning, Medicine and the 'Placebo Effect' by Moerman DE, Cambridge University Press 2002, p.34, summarising secondary references to five further studies

81 'one of the most impressive': Moerman DE, Harrington A. Making space for the placebo effect in pain medicine. Sem in Pain Med (March 2005); 3 (1 Spec issue): 2–6

81 'Another study from 2002': Walsh BT, Seidman SN, Sysko R, Gould M. Placebo response in studies of major depression: variable, substantial, and growing. JAMA (10 April 2002); 287 (14): 1840–7

84 'one study found': Ernst E, Schmidt K. Aspects of MMR. BMJ (2002); 325: 597

Chapter 6: The Nonsense *du Jour*

89 'It is impossible': Harry G. Frankfurt, *On Bullshit*. Princeton University Press (2005) http://press.princeton.edu/video/frankfurt

105 'over $50 billion': http://www.nutraingredients-usa.com/news/ng.asp?n=85087

105 'One was in Finland': Alpha-Tocopherol Beta-Carotene Cancer Prevention Study Group. The effect of vitamin E and beta carotene on the incidence of lung and other cancers in male smokers. New Eng J Med (1994); 330: 1029–35

106 'Two groups of people': Thornquist MD, Omenn GS, Goodman GE, Grizzle JE, Rosenstock L, Barnhart S, Anderson GL, Hammar S, Balmes J, Cherniack M. Statistical design and monitoring of the Carotene and Retinol Efficacy Trial (CARET). Control Clin Trials (1993); 14: 308–24; Omenn GS, Goodman GE, Thornquist MD, Balmes J, Cullen MR, Glass A, et al. Effects of a combination of beta carotene and vitamin A on lung cancer and cardiovascular disease. N Engl J Med (1996); 334: 1150–5.http://jnci.oxfordjournals.org/cgi/ijlink?linkType=ABST&journalCode=nejm&resid=334/18/1150]

106 'The most up-to-date': Vivekananthan DP et al. Use of antioxidant vitamins for the prevention of cardiovascular disease: meta-analysis of randomised trials. Lancet (2003); 361: 2017–23 http://www.thelancet.com/journals/lancet/article/PIIS0140673603136379/abstract

107 'The Cochrane review': Caraballoso M, Sacristan M, Serra C, Bonfill X. Drugs for preventing lung cancer in healthy people. Cochrane Database of Systematic Reviews (2003); 2

107 'a Cochrane review': Bjelakovic G, Nikolova D, Gluud LL, Simonetti RG, Gluud C. Antioxidant supplements for prevention of mortality in healthy participants and patients with various diseases. Cochrane Database of Systematic Reviews (2008); 2

108 'Dr Benjamin Spock': Chalmers I. Invalid health information is potentially lethal. BMJ (2001); 322 (7292): 998

109 'price-fixing cartel': John M. Connor Kluwer, *Global Pricefixing: Our Customers Are the Enemy*. Springer (2001). Available online: http://books.google.co.uk/books?id=7M8n4UN23WsC

Notes

109 'Doubt is our product': David Michaels (ed.) *Doubt is Their Product: How Industry's Assault on Science Threatens Your Health.* Oxford University Press (2008)

Chapter 7: Dr Gillian McKeith PhD

114 'Dudley J. LeBlanc': Ann Anderson, *Snake Oil, Hustlers and Hambones: The American Medicine Show.* McFarland (2005)

121 'During the war': Commencement Speech from Caltech 1974, also in Richard Feynman. *Surely You're Joking, Mr. Feynman!: Adventures of a Curious Character.* WW Norton (1985)

123 Clayton College of Natural Health website: http://www.ccnh.edu/about/programs/tuition.aspx

Chapter 8: 'Pill Solves Complex Social Problem'

136 'a large, well-conducted': Hutchings J, Bywater T, Daley D, Gardner F, Whitaker C, Jones K, Eames C, Edwards RT. Parenting intervention in Sure Start services for children at risk of developing conduct disorder: pragmatic randomised controlled trial. BMJ (2007); 334: 678

136 'cost-effectiveness analysis': Edwards RT, Ó Céilleachair A, Bywater T, Hughes DA, Hutchings J. Parenting programme for parents of children at risk of developing conduct disorder: cost effectiveness analysis. BMJ (2007); 334: 682

146 'one study by a researcher': Richardson AJ, Montgomery P. The Oxford-Durham study: a randomized, controlled trial of dietary supplementation with fatty acids in children with developmental coordination disorder. Pediatrics (2005); 115 (5): 1360–6

153 'disease-mongering': Moynihan R, Doran E, Henry D Disease mongering is now part of the global health debate. PLoS Med (2008); 5 (5): e106. doi:10.1371/journal.pmed.0050106 A good place to start your reading on disease-mongering

159 'Professor Hywel Williams': Williams HC. Evening primrose oil for atopic dermatitis. BMJ (2003); 327: 1358–9

160 'the most popular food supplement product in the UK': 'The four markets dominating EU supplements' http://www.nutraingredients-usa.com/news/ng.asp?n=85087; "Galenica assumes control of Equazen Nutraceuticals based in

331

the UK". Press release. http://www.galenica.com/Galenica/en/
archive/media/releases/2006_12_04_21398644_meldung.php

Chapter 9: Professor Patrick Holford

164 'He has claimed that he has corrected his book': Professor Holford
did not change the main text in the book chapter. He added some text to the
note at the back, in a small font, referencing some other papers where people
did actually, at least, tip both AZT and vitamin C onto cells in a dish (this
changes nothing), and a demand for more research, none of which, I note, he
has offered to fund from his own extensive corporate involvement in this $50
billion sector. He is, after all, Head of Science and Education at the food
supplement pill company BioCare, which sells a vitamin C pill in tubs with his
face on. To be fair, he did however once deliver my favourite line from five years
of writing in this area: 'Perhaps Goldacre, who purports to be the campaigner
for evidence-based medicine, could provide some evidence that high-dose
vitamin C has no effect against HIV AIDS.'

166 'systematic review from Cochrane': Douglas RM, Hemilä H,
Chalker E, Treacy B. Vitamin C for preventing and treating the
common cold. Cochrane Database of Systematic Reviews
(1998); 1. Date of last update: 14 May 2007 (Cochrane reviews
are constantly updated, and all previous versions are available,
so you can see what they said at various times in the past too)

167 'Dr Richard Smith': Smith R. Investigating the previous studies
of a fraudulent author. BMJ (2005); 331: 288–291; Hamblin T.
The Secret Life of Dr Chandra. BMJ (2006); 332: 369

167 'three-part investigative documentary series': The documentary
on Dr Chandra can be watched online at:
http://www.cbc.ca/national/ news/chandra/

167 'a retrospective re-analysis': Hemilä H, Herman ZS. Vitamin C
and the common cold: a retrospective analysis of Chalmers'
review. J Am Coll Nutr (April 1995); 14 (2): 116–23.

169 'a systematic review and meta-analysis': Vivekananthan DP et
al. Use of antioxidant vitamins for the prevention of
cardiovascular disease: meta-analysis of randomised trials.
Lancet (2003); 361: 2017–23

178 'the QAA's report':
http://www.qaa.ac.uk/reviews/reports/institutional/Luton1105/
RG162UniLuton.pdf

Chapter 10: Is Mainstream Medicine Evil?

182 'From the state of current knowledge':
http://clinicalevidence.bmj.com/ceweb/about/knowledge.jsp

182 'These real-world studies': The classic general medicine
reference for this is Ellis J, Mulligan I, Rowe J, Sackett DL.
Inpatient general medicine is evidence based. A-Team, Nuffield
Department of Clinical Medicine. Lancet (12 August 1995); 346
(8972): 407–10. There have been numerous copycat studies in
various specialities, and rather than list them here, an excellent
review of them is maintained at http://www.shef.ac.uk/scharr/
ir/percent.html

183 'all of those studies': Mayor S. Audit identifies the most read
BMJ research papers. BMJ (2007); 334: 554–5; Hippisley-Cox J,
Coupland C. Risk of myocardial infarction in patients taking
cyclo-oxygenase-2 inhibitors or conventional non-steroidal
anti inflammatory drugs: population based nested case-control
analysis. BMJ (2005); 330: 1366; Gunnell J, Saperia J, Ashby D.
Selective serotonin reuptake inhibitors (SSRIs) and suicide in
adults: meta-analysis of drug company data from placebo
controlled, randomised controlled trials submitted to the
MHRA's safety review. BMJ (2005); 330: 385; Fergusson D et al.
Association between suicide attempts and selective serotonin
reuptake inhibitors: systematic review of randomised controlled
trials. BMJ (2005); 330: 396

187 'whole areas can be orphaned': Iribarne A. Orphan diseases and
adoptive initiatives. JAMA (2003); 290: 116; Francisco A. Drug
development for neglected diseases. Lancet (2002); 360: 1102

190 'If you follow the references': Safer DJ. Design and reporting
modifications in industry-sponsored comparative
psychopharmacology trials. J Nerv Ment Dis (2002); 190: 583–92

190 'various studies have shown': Modell et al. (1997); Montejo-
Gonzalez et al. (1997); Zajecka et al. (1999); Preskorn (1997): in
Safer, ibid.

193 'If the difference': Pocock SJ. When (not) to stop a clinical trial
for benefit. JAMA (2005); 294: 2228–30

194 'a systematic review': Lexchin J, Bero LA, Djulbegovic B, Clark
O. Pharmaceutical industry sponsorship and research outcome
and quality. BMJ (2003) 326: 1167–70

194 'One review of bias': Rochon PA, Gurwitz JH, Simms RW, Fortin PR, Felson DT, Minaker KL, Chalmers TC. A study of manufacturer-supported trials of nonsteroidal anti-inflammatory drugs in the treatment of arthritis. Arch Intern Med. (24 January 1994); 154 (2): 157–63

194 'when the methodological flaws': Lexchin J, Bero LA, Djulbegovic B, Clark O. Pharmaceutical industry sponsorship and research outcome and quality: systematic review. BMJ (31 May 2003); 326 (7400): 1167–70

196 'In 1995, only': Schmidt K, Pittler MH, Ernst E. Bias in alternative medicine is still rife but is diminishing. BMJ (3 November 2001); 323 (7320): 1071

196 'A review in 1998': Vickers A, Goyal N, Harland R, Rees R. Do certain countries produce only positive results? A systematic review of controlled trials. Control Clin Trials (April 1998); 19 (2): 159–66

198 'a paper has even found': Dubben H, Beck-Bornholdt H. Systematic review of publication bias in studies on publication bias. BMJ (2005); 331: 433–4

198 'published a paper': Turner EH, Matthews AM, Linardatos E, Tell RA, Rosenthal R. Selective publication of antidepressant trials and its influence on apparent efficacy. N Eng J Med (17 January 2008); 358 (3): 252–60

199 'A classic piece of detective work': Tramer MR, Reynolds DJM, Moore RA, McQuay, H J. Impact of covert duplicate publication on meta-analysis: a case study. BMJ (1997) 315: 635–40

200 'When we carried out': Cowley AJ et al. Int Journ Card (1993) 40: 161–6

201 'the three highest-ranking papers': Audit identifies the most read BMJ research papers. BMJ (17 March 2007); 334: 554–5

202 'It is a shame': Scolnick EM. Email communication to Deborah Shapiro, Alise Reicin and Alan Nies re: Vigor. 9 March 2000. http://www.vioxxdocuments.com/Documents/Krumholz_Viox x/Scolnick2000.pdf]

202 'The *New England Journal of Medicine*': Curfman GD, Morrissey S, Drazen JM. Expression of concern reaffirmed. NEJM (16 March 2006); 354 (11):1193

203 'a US company': Gottlieb S. Firm tried to block report on failure of AIDS vaccine. BMJ (2000); 321: 1173

203 'The drug company': Nathan D, Weatherall D. Academia and industry: lessons from the unfortunate events in Toronto. Lancet; 353; 9155: 771–2

205 'These adverts have been': Gilbody et al. Benefits and harms of direct to consumer advertising: a systematic review. Qual Saf Health Care (2005); 14: 246–50
http://qshc.bmj.com/cgi/content/full/14/4/246

Chapter 11: How the Media Promote the Public Misunderstanding of Science

211 'Nick Davies': Davies N. *Flat Earth News*. Chatto & Windus (2008)

218 'Research on smoking': Proctor RN. Schairer and Schöniger's forgotten tobacco epidemiology and the Nazi quest for racial purity. Int. J. Epidemiol 30: 31–4

219 'John Ioannidis': Ioannidis JPA. Why most published research findings are false. PLoS Med (2005) 2 (8): e124

Chapter 12: Why Clever People Believe Stupid Things

227 'a classic experiment': Gilovich T, Vallone R, Tversky, A. The hot hand in basketball: on the misperception of random sequences. Cog Psych (1985); 17: 295–314

229 'ingeniously pared-down experiment': Schaffner PE. Specious learning about reward and punishment. J Pers Soc Psych (June 1985); 48 (6): 1377–86

231 'In one experiment': Snyder M, Cantor N. Testing hypotheses about other people: the use of historical knowledge, J Exp Soc Psych (1979); 15: 330–42

232 'The classic demonstration': Lord CG, Ross L, Lepper MR. Biased assimilation and attitude polarisation: the effects of prior theories on subsequently considered evidence. J Pers Soc Psyc (1979); 37: 2098–109

234 'In one, subjects': Tversky A, Kahneman D. Availability: a heuristic for judging frequency and probability. Cog Psych (1973), 5: 207–32

236 'Asch's experiments': Asch SE. Opinions and social pressure. Sci Am (1955); 193: 31–5

237 'the behaviour of sporting teams': Frank MG, Gilovich T. The dark side of self- and social-perception: black uniforms and aggression in professional sports. J Pers Soc Psych (January 1988); 54 (1): 74–85

238 'It's not safe': The experiments in this chapter, and many more, can be found in *Irrationality* by Stuart Sutherland and *How We Know What Isn't So* by Thomas Gilovich

Chapter 13: Bad Stats

239 'Let's say the risk': Gigerenzer G. *Reckoning with Risk*. Penguin (2003)

240 'Natural frequencies': Butterworth et al. Statistics: what seems natural? Science (4 May 2001): 853

240 'The other methods': Hoffrage U, Lindsey S, Hertwig R, Gigerenzer G. Communicating statistical information. Science (22 December 2000); 290 (5500): 2261–2

241 'there are studies': Hoffrage U, Gigerenzer G. Using natural frequencies to improve diagnostic inferences. Acad Med (1998); 73: 538–40

253 'the same test': Gigerenzer G. *Adaptive Thinking: Rationality in the Real World*. Oxford University Press (2000)

253 'Let's think about': Szmukler G. Risk assessment: 'numbers' and 'values'. Psych Bull (2003) 27: 205–7

257 'a small collection': www.qurl.com/lucia

Chapter 14: Health Scares

262 'An academic paper': Manning N, Wilson AP, Ridgway GL. Isolation of MRSA from communal areas in a teaching hospital. J Hosp Infect (March 2004); 56 (3): 250–1

268 'Kruger and Dunning': Kruger J, Dunning D. Unskilled and unaware of it: how difficulties in recognizing one's own incompetence lead to inflated self-assessments'. J Pers Soc Psych (1999); 77; 6: 121–34

269 'In 1957, a baby': Brynner R, Stephens TD. *Dark Remedy: The Impact of Thalidomide and its Revival as a Vital Medicine*. Perseus Books (2001)

270 'Philip Knightley': Excerpted in Pilger J (ed.). *Tell me no Lies*. Cape (2004)

270 'Many years later': Thalidomide hero found guilty of scientific fraud. New Scientist (27 February 1993)

Chapter 15: The Media's MMR Hoax

278 '12 children': Wakefield AJ, Murch SH, Anthony A et al. Ileal-lymphoid-nodular hyperplasia, non-specific colitis, and pervasive developmental disorder in children. Lancet (1998); 351 (9103): 637–41

280 'one of the few': e.g. Chess S. Autism in children with congenital rubella. J Autism Child Schizophr (January–March 1971); 1 (1): 33–47

280 'it is investigating': http://briandeer.com/wakefield/wakefield-deal.htm

282 'including the BBC': No jabs, no school says labour MP. http://news.bbc.co.uk/1/hi/health/7392510.stm

288 'one survey': Schmidt K, Ernst E, Andrews. Survey shows that some homoeopaths and chiropractors advise against MMR. BMJ (14 September 2002); 325 (7364): 597

288 'Thirty-two per cent': Hargreaves I, Lewis J, Speers T. Towards a better map: science, the public and the media, Economic and Social Research Council (2003). http://www.esrc.ac.uk/ESRCInfoCentre/Images/Mapdocfinal_tcm6–5505.pdf

289 'peak of the media': Boyce T. *Health, Risk and News: The MMR Vaccine and the Media*. Peter Lang Publishing Inc. (2007)

290 'published a paper': Ibid.

291 '*not a single one*': Durant J, Lindsey N. GM foods and the media. Select Committee on Science and Technology, Third Report, Appendix 5 www.publications.parliament.uk/pa/ld199900/ldselect/ldsctech/38/3810.htm

295 'a systematic review': Smeeth L, Cook C, Fombonne E, Heavey L, Rodrigues LC, Smith PG et al. MMR vaccination and pervasive developmental disorders: a case-control study. Lancet (2004); 364 (9438): 963–9

297 'This study was big': Madsen KM, Hviid A, Vestergaard M, Schendel D, Wohlfahrt J, Thorsen P et al. A population-based study of measles, mumps, and rubella vaccination and autism. N Eng J Med (2002); 347 (19): 1477–82

300 'Scientists in America': http://www.telegraph.co.uk/news/
      main.jhtml?xml=/news/2002/06/23/nmmr23.xml
302 'a very similar study': Afzal MA, Ozoemena LC, O'Hare A et al.
      Absence of detectable measles virus genome sequence in blood
      of autistic children who have had their MMR vaccination
      during the routine childhood immunization schedule of UK. J
      Med Virology (2006); 78; 5: 623–30
303 'Another major paper': D'Souza, Y, et al. No evidence of
      persisting measles virus in peripheral blood mononuclear cells
      from children with autism spectrum disorder. Pediatrics (4
      October 2006); 118: 1664–75
306 'In some parts': http://www.westminster-
      pct.nhs.uk/news/mmr0405.htm; Pearce et al. Factors associated
      with uptake of measles, mumps, and rubella vaccine (MMR)
      and use of single antigen vaccines in a contemporary UK
      cohort: prospective cohort study. BMJ (2008); 336 (7647): 754
307 'a systematic review': Chapman S et al. Med J Aust. (5
      September 2005); 183 (5): 247–50. Grilli R et al. Cochrane
      Database of Systematic Reviews (2001); 4: CD000389
307 'A mischievous paper': Phillips DP et al. N Engl J Med (1991);
      325: 1180–3
308 'systematic quantitative surveys': Schwitzer. G PLoS Med
      (2008); 5 (5): e95
308 'Meanwhile, the incidence': HPA. Confirmed measles mumps
      and rubella cases in 2007: England and Wales. Health
      Protection Report (2008); cited 9 April 2008; 2 (8).
      http://www.hpa.org.uk/hpr/archives/2008/hpr0808.pdf
309 'Congenital rubella syndrome': Fitzpatrick M. MMR: risk,
      choice, chance. Brit Med Bulletin (2004); 69: 143–53
312 'epidemic in 2005': Gupta RK, Best J, MacMahon E. Mumps
      and the UK epidemic. BMJ (14 May 2005); 330: 1132–5

And Another Thing
319 'The true cost':
      http://www.economist.com/research/Economics/alphabetic.cfm
      ?letter=O